AF382497

I am very grateful to my wife Petya and my daughters Sofia and Victoria - for their patience with me while working on this project. There have been many late nights, and much missed time with family, along the way. Thank you for supporting me!

Thanks to all the reviewers of the book and the I-CM - especially Prof. Marcos Esterman from RIT, Emil Echkov from Visteon Corporation, and Stefan Vlaev from Intellics Engineering.

SYSTEMS ENGINEERING FOR ALL

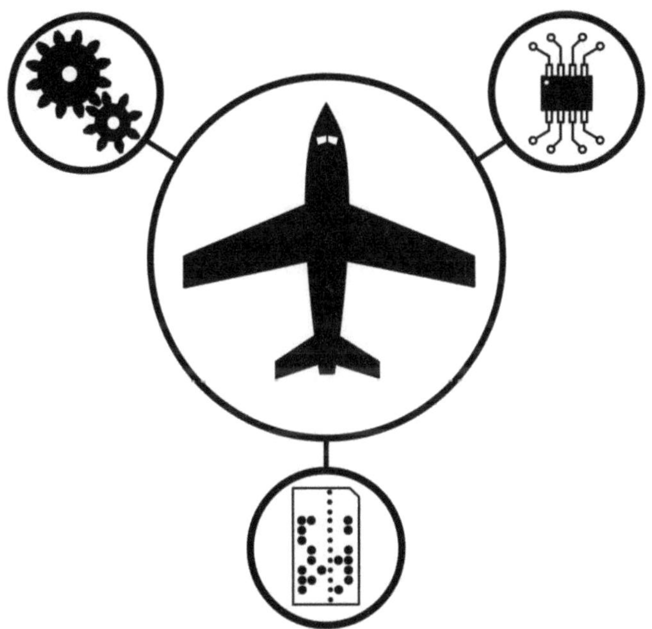

Introduction to Systems Engineering for non-systems engineers.

ISBN: 978-3-347-10356-6 (Paperback)
ISBN: 978-3-347-10357-3 (Hardcover)
ISBN: 978-3-347-10358-0 (e-Book)

Verlag und Druck: tredition GmbH, Halenreie 40-44, 22359 Hamburg

This book is a quick, practical introduction to systems engineering for non-systems engineers. It is written in a popular form and requires only basic engineering knowledge. It uses diverse examples to ensure coverage of various disciplines and industry cases.
*The book includes an introduction to the minimalistic Interface-Component Model (**I-CM**) for systems engineering, which may be used to support the activities discussed in the book. However, the I-CM and the book may be used independently from each other. The I-CM method is the author's own development.*
To find all about I-CM (including free downloads), read articles, and follow book updates - visit:
www.systems-engineering.org

This book is a result of the cumulative knowledge of the author and its information may be traced to many sources - literature, lectures, practice, industry standards, etc. The References section at the end has a list of excellent books and publications which the author has used in developing this book.
For detailed and officially recognized practices and methods in systems engineering, the reader may refer to "INCOSE Systems Engineering Handbook."

The trademarks and copyrights are marked to the best of the author's knowledge.
The author does not bear responsibility for errors or omissions, nor any liability of using this book or the I-CM tool.
The advice and conclusions are solely the author's own understanding and opinion.

CONTENTS:

Modern Engineering	7
Introduction to Systems	12
Special Attention to the Interfaces	15
Introduction to Systems Engineering	20
Describe the System (Requirements Definition)	24
Design the Solution (System Architecture)	40
Integrate the Parts (System Integration)	57
Validate the Result (System Testing)	80
Introduction to the I-CM Method	86
Quick Reference to I-CM 1.x	98
Abbreviations	110
Bibliography	112

Modern Engineering

Imagine a toothbrush...

An unspectacular daily object, which we use without even thinking about, as it has been around for ages (Figure 1).

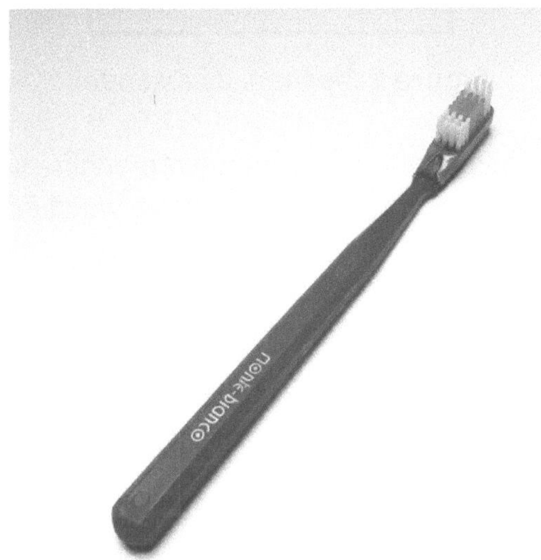

Figure 1 *A toothbrush*

To industrially engineer a classic toothbrush, we may need a mechanical engineer (Figure 2) who can do the drawings and design the toolings for its production.

Figure 2 *Mechanical engineer*

In the past, the toothbrushes were not even engineered, but simply made by hand with natural materials such as wood and horsehair.

This is a toothbrush, too (Figure 3). It has the same main function as the classic toothbrush, but different performance characteristics:

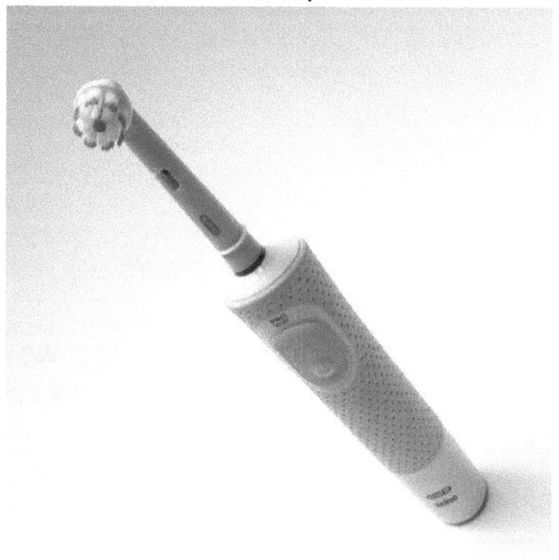

Figure 3 *Electrical toothbrush*

To design this one as an industrial product, we will need a multidisciplinary team of mechanical, electrical, and software engineers (Figure 4) to create:

- A relatively complex mechanical construction of a water-resistant body, which houses sensitive electro-mechanical and electronics assemblies and supports detachable brushes
- A drivetrain with an electrical motor, allowing mechanical oscillation of the attached brush
- An electronics module (usually with a microcontroller), which controls the motor, the operational modes of the brush, the timer, and the battery charging circuit
- Software to drive the microcontroller
- A power source (usually a battery)
- And a movable brush assembly which fits the toothbrush body

Figure 4 *ME, SW and EE engineers*

The simple daily object has evolved into a complex engineering system, which needs to be designed and managed. The rise of complexity demonstrated with the toothbrush can be seen practically everywhere today - once mechanical systems turn to mechatronics, discrete electronics solutions are now driven by tiny microcontrollers with software running on them, and personal computing is now distributed over gigantic planet-scale infrastructure. But how to engineer such complex multidisciplinary systems? Is there someone (Figure 5) who designs and manages the system as a whole, above the specialists which work with its parts only?

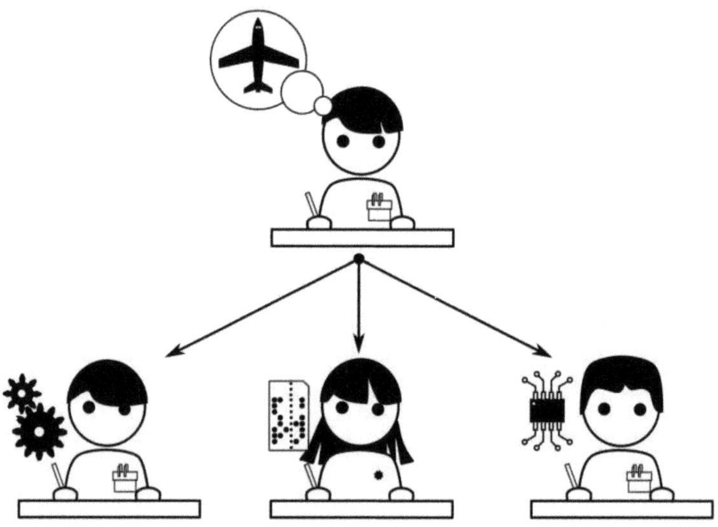

Figure 5 *Multidisciplinary team orchestrated by a SE*

> *Modern engineered products are complex and multidisciplinary.*

Systems engineering (SE) is the "interdisciplinary discipline" which aims at mastering the system's complexity. As dark matter keeps the universe from falling apart, systems engineering unites the different

10

engineering disciplines to deliver an integral solution to the specific need.

Technical systems with high complexity are not a new phenomenon - airplanes, ships, and automobiles have been around for some time. In those industries, systems engineering is well known and respected - usually experienced engineers with natural systems thinking and a holistic engineering approach are promoted to manage the complexity of those large-scale multidisciplinary products. However, the recently increased complexity is a deadly trap for many other industries, which organizations don't know how to handle.

This book is a short practical introduction to systems engineering for non-systems engineers. A minimalistic SE tool, called the I-CM (Interface-Component Model), is used as a companion throughout the book and allows managing all the discussed SE activities in one place. The I-CM was invented together with the book, but it is a stand-alone method and tool, which can be used in practice independently of the book. The same applies to the book - it doesn't necessarily require the I-CM to be understood.

Introduction to Systems

A system is a group of things collaborating to achieve a common goal. An important property of a system is that it delivers more value than the sum of its parts. Where does this additional value come from? It comes from the interrelations between the parts!
Examples of systems range from a ballpoint pen to a spaceship, and from molecules to the universe. A football team is a system consisting of 11 players who interact with each other to deliver higher performance than 11 individuals playing solo.

> *Engineered systems are the focus of this book.*

In engineering, a system usually consists of components interacting via interfaces, which together satisfy a set of requirements and goals. Let's consider the following example of an engineering system as a solution satisfying a set of requirements.

A vehicle is a thing used to transport people or goods. If we refine the requirements to a land-vehicle, this thing needs to transport the objects by moving them on a trajectory parallel to the landscape, while keeping the transported objects motionless related to the vehicle. Most vehicles use wheels (due to their rolling properties) to move on land. But a wheel alone does not cover all the requirements - especially the trajectory part. If we fix the transported object to a simple wheel, it will follow a trajectory like in Figure 6 (in case of a passenger, this wouldn't be the most pleasant ride):

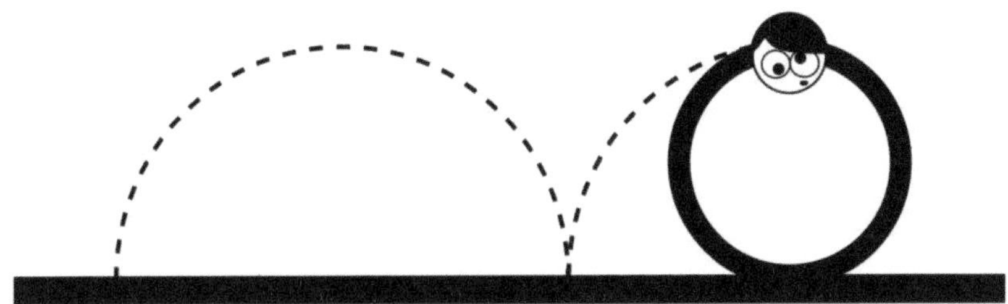

Figure 6 *A wheel as a vehicle*

If we place the object on a board, however, it will remain motionless. Unfortunately it will be motionless not only related to the board, but also related to the environment - in other words, it will not be transported at all (Figure 7).

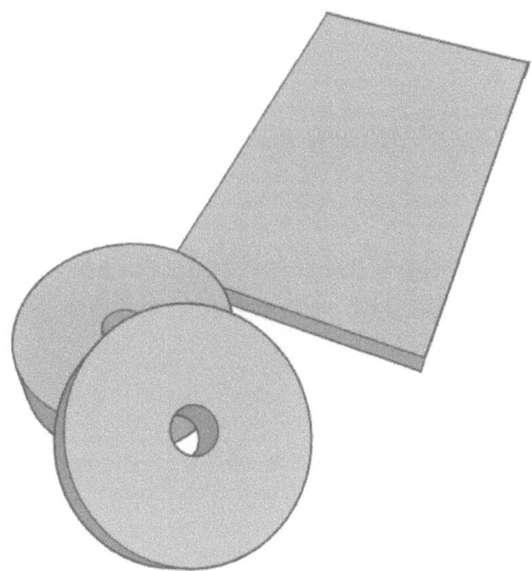

Figure 7 *Parts to build a vehicle*

But what if we make these components interact in a specific way, called "bearing"? Then we get a system which will have the function of *transportation* - a feature not present in either the wheel or the board (Figure 8).

Figure 8 *Simple vehicle*

The interaction bearing is the **interface** between the wheel and the board, constraining the motion in the desired direction only.

Special Attention to the Interfaces

Interfaces tend to get much less attention than components from many engineers. Educated systems engineers, however, usually have the topic of "Systems Thinking" as a foundation in their education programs, and pay more attention to interfaces.

A very important consideration is that an interface in the architecture may later become a subsystem composed of both physical components and other interfaces! We will illustrate this with the vehicle example from above.

Looking at the picture of our super simple Vehicle, it seems that there is something more than just the board and the wheels - the axle. Then, is the axle a component we haven't considered while assembling our Vehicle system? Depends - below are two examples, where the axle is either a separate component or a function of the shape of the "board" component (Figure 9).

Figure 9 *Two vehicle models - axis as feature, and as part.*

It is important to understand that these two Vehicles have the same architecture, but different designs. As you can see, the implementation of the interface can be propagated to the physical components in different ways, depending on the design decision. The design decisions for the interface implementation are often driven by specific requirements, which can be decomposed to all elements of the subsystem - the interface and the components. For example, a simple, generic bearing implementation is the plain bearing, where the axle fits in a hole in the center of the wheel, as shown in Figure 10.

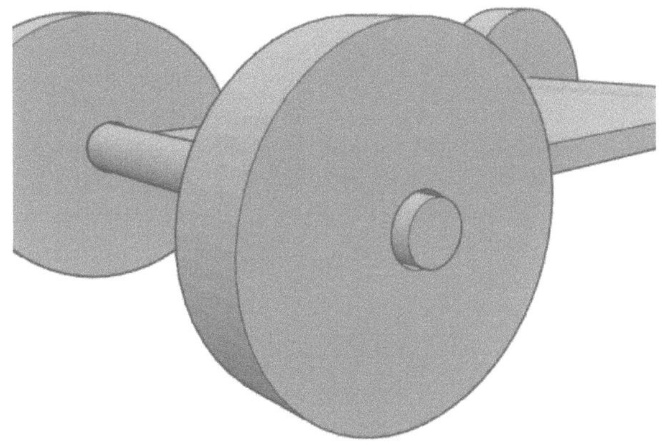

Figure 10 *Plain bearing*

This design, while simple and effective, is not long lasting and produces heat with fast rotation due to the high level of friction and wear between the axle and the wheel. This friction is not a direct result of the components themselves, but of the *bearing* interaction between them - having the same assembly in a static (not moveable) setup will not produce the wear, but will also not deliver the *transport* function.

Let's add additional requirements to the bearing interface - to be reliable, long-lasting and support higher speeds. The interface could then be implemented as shown in Figure 11.

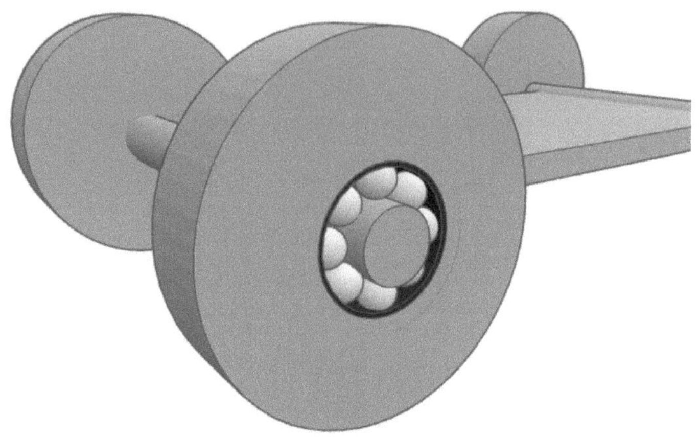

Figure 11 *Ball bearing*

Using a ball bearing, we can cover these requirements by translating the mechanical friction to rolling. But we see some new physical components and interfaces (a subsystem) are introduced into the system. This subsystem establishes new (now static) interfaces to the axle and to the wheel. And still, on an architectural level, this implementation doesn't differ from the previous solution - it is a design decision of how to implement the bearing interface.

Thanks to the interactions, the system delivers more than the sum of its parts!

I hope this example shows how multilayered the interfaces are. Although they are "only" interactions between physical components, interfaces can have physical, logical, and performance characteristics, can propagate new interfaces and components downstream in the design process, etc. Interfaces need ownership, design, and verification, just like components.

The introduction of a subsystem within the interface, which the ball bearing example shows, may raise the question - when can we say that the parts of a system are really parts and not other systems? What is the difference between a system (of parts) and System of Systems (SoS)? My own conviction is that there is no difference, as the definition of a part (as something atomic) is purely a subjective decision depending on the project context. If we are electronics engineers designing a PC hardware, the CPU is a part for us. But not for the CPU designers - for them it is an extremely complex system. The "System of Systems" is only the way we look at the system - as soon as we have more than one level of hierarchy in the system decomposition, the system may be seen as a System of Systems. *However, here my opinion differs from some officially accepted standards, like ISO 21840! To be compliant with the standards, please refer to the definitions there.*

> *Almost every system is actually a "System of Systems."*

Introduction to Systems Engineering

Systems Engineering (SE) is a complex interdisciplinary approach to define, design, analyze and manage complex systems and their lifecycles. In this book, we will highlight a subset of systems engineering which focuses on the development of engineered systems (not natural or social ones, like the Solar System or a tribe). In this book, I will propose a minimal set of four activities, to establish a valuable systems engineering approach:
- Requirements Definition
- System Architecture
- System Integration
- System Testing

Even if we refer to some industry standards (like ASPICE), we may still simplify their defined system processes down to these four as well. *Note: Implementing the activities as described here is not enough to make the organization ASPICE compliant!*

> *Systems engineering is a holistic approach to engineer complex multidisciplinary systems.*

The requirements define *what* our system should do (behave, perform, etc.). They describe the problem and the targets, but not the solution.

The architecture is a description of the physical solution to a problem. It defines *how* we satisfy the requirements. The architecture usually also decomposes the solution (system) to its parts and interactions (components and interfaces), which are propagated downstream for detailed design and implementation.

The integration activity should put the designed parts together and ensure their correct interactions. During integration, the system should be verified against the architecture to ensure that all components deliver the expected performance and interact with each other in the expected way.

The testing activity should verify that the final integrated system satisfies the requirements (serves the purpose). Testing includes verification and validation activities, which ensures that "the system does the right things in the right way."

These activities (in blue) are represented in the engineering flow in Figure 12.

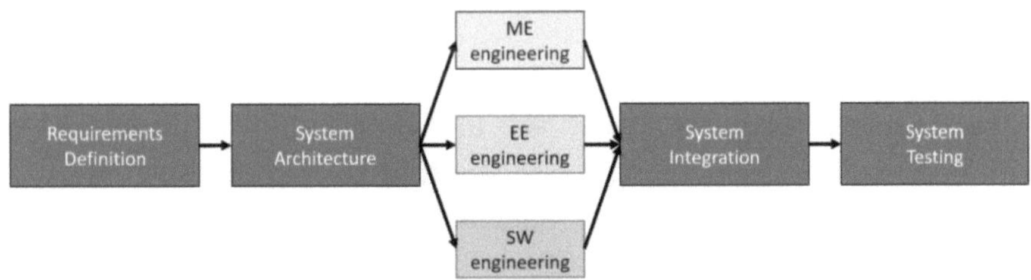

Figure 12 *Linear abstraction view of an engineering workflow*

Sometimes this process is presented like the letter "V" (referred to as the V-Model, adapted here to the book's scope, and shown in Figure 12A). The line of the "V" from left to right shows the same process sequence as in Figure 12. On this representation, however, the process input dependency is clear (the opposite sides of the "V" on the same level - see the "satisfies" arrows). It is also grouped such that the left side of the "V" is dedicated to analysis and design, and the right side to assembly and verification.

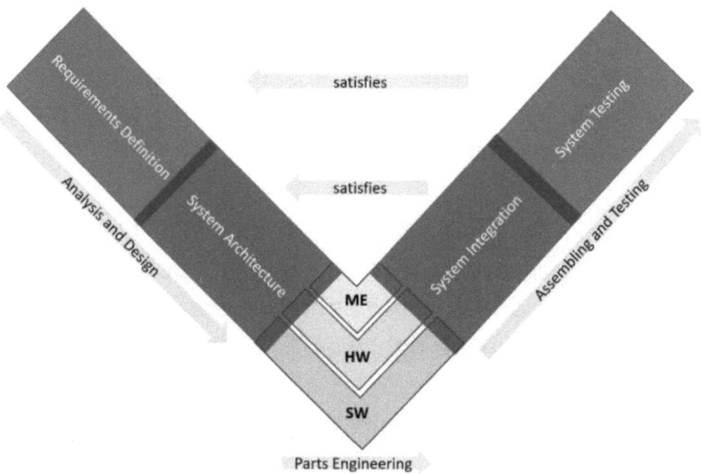

Figure 12A *Simple "V"-type process model*

In the next four chapters, we will discuss in more detail what these systems engineering activities are and how to perform them. After that, we will introduce the I-CM method and show how we can implement the basics of these activities in a very simple manner.

The systems engineer works with the whole.

But before we move further, let's say a few words about the people behind the systems engineering - the systems engineers. Usually, we speak about engineers as *"specialists."* A specialist has a domain of competence and mastery at a high level of detail. The systems engineer, however, is a *"generalist."* The systems engineer should think holistically, comprehend the whole, and understand the parts and the relations. The generalist steers the specialists on the "whole" dimension, while the specialists work on the "details" dimension (Figure 13).

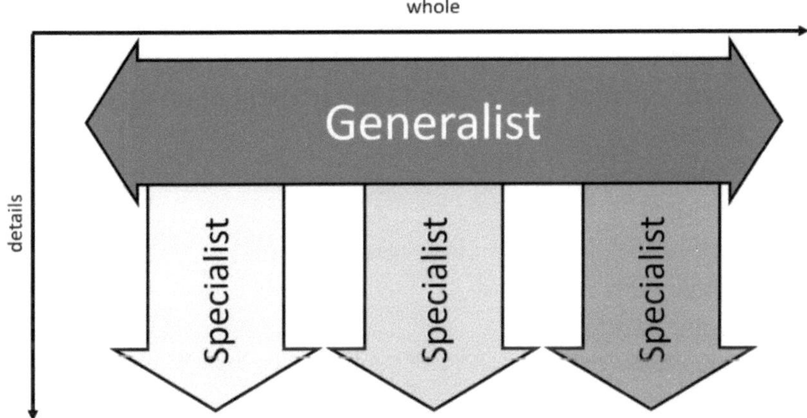

Figure 13 *Generalist and specialists distribution*

In most cases, systems engineers are former specialists (already-educated engineers in one of the classical disciplines) with enough experience and systems thinking. In addition, many universities and institutions provide educational programs and academic degrees in systems engineering.

Describe the System (Requirements Definition)

Before we start developing something, we need to understand what it shall be. It needs to serve a purpose, solve a problem, or fill a gap. In other words - we need to describe what we want to create, by whom, where, how, and for what it will be used. This description will serve as a basis for the engineering activities, and the requirements definition is where this happens.

The requirements specification is a set of requirements. A requirement is a statement that describes a function, behavior, or property; sets a constraint; gives a performance target; etc.
In the literature, criteria for a "good" requirement often includes:
- Atomic
- Feasible
- Testable
- Non-redundant (with other requirements)
- Solution-neutral
- Understandable
- Objective

In addition, I have observed widespread usage of "shall" and "must" for absolutely mandatory requirements, and "may" and "should" for recommended or optional ones. It is not a universal standard and may differ between industries and regions. A good idea is to establish such a dictionary for the organization at the beginning.

In this chapter, we will see how to come to the requirements specification and whether the requirements shall always fulfill all the criteria above.

Requirements engineering is a complex and creative activity and includes:
- Identifying a need
- Stating the problem
- Generating concept ideas
- Specifying the system

Identifying a need is like, "If you are looking for trouble..." - find some aspect of the life of a person, group of people, animals, plants, the planet (you name it), which could be made better. For example - human transportation. Let's assume that we have found the need to transport people from point A to point B. We don't know the solution for it, that's why it is a big (or maybe not that big) black box for us. And the need statement is quite solution neutral: "Transport person from point A to point B" (Figure 14).

Figure 14 *Problem statement*

It sounds like we found our need, and even kind of stated the problem. But in this way we will never find any real solution, as the problem is too generic and the solutions to such problems are too abstract - we can leave the person walking, we may use a rocket, airplane, car, elevator, ship, bike… a teleporter maybe (Figure 15)?

Figure 15 *Potential general solutions without constraints*

This is because we have too many questions: What are those points A and B - do we want to transport the person to the grocery store and back, from Paris to London, from Europe to North America, from Earth to the Moon? Can we accept a solution which costs a billion to buy? *Even - Why do we need to transport the person in the first place? Is the person's presence at Point B really necessary or could it be arranged differently?*
Stating the problem should include all the relevant aspects around our system, including all relevant stakeholders, constraints, interfaces, environments, and even critical use cases.

In other words, we need to define the *boundaries* of our system, and the *context* in which it will exist (Figure 16).

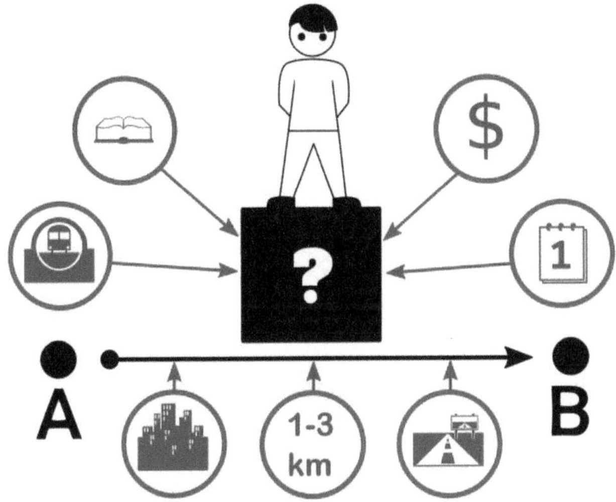

Figure 16 *Adding context and constraints*

A system has boundaries and relevant context.

Now, we have much more clarity about our system and its context. Let's try to describe the diagram above, and let's assume that our system is now called a "Vehicle":
- *The Vehicle shall be a transportation device for a single person*
- *The Vehicle shall be human-powered*
- *The end-price of the Vehicle should be less than 1000 EUR*

- *The Vehicle must be legally- and socially-acceptable*
- *The technology of the Vehicle should be currently available and common*
- *The Vehicle shall be compact and able to be carried on public transportation*
- *The environment of the Vehicle will be city/urban*
- *The Vehicle is supposed to be used on a road/pavement, regulated for man-powered vehicles (bike roads) or common roads when allowed*
- *The optimal distance for the usage of the Vehicle is assumed to be 1 (one) to 3 (three) km*
- *The Vehicle should be at least three times faster than a regular walk (on a flat surface)*

While the requirements are still solution neutral, we have more clues about what could be "inside" our black box. The more constraints, performance targets, and specific use cases we add, the more concrete and less abstract our specification will be. However, many of the requirements above are still meta- or implicit-requirements. Let's take the following example:
- *The Vehicle should be compact and able to be carried on public transportation*

This is kind of a user requirement, but not a system requirement. "Compact" is very subjective, and being "able" to take some object on public transportation varies between regions and types of transport. Out of this requirement, there should be a set of derived requirements that define more concrete goals. Further refinement of the context may lead us to assume public city transportation (e.g., bus, tram, metro) over the main target markets (e.g., Europe, USA, Canada, Australia), and the common allowance to transport goods (volume, size, and weight). All these will generate more specific targets, like:

- *Weight up to 18 kg*
- *Carry size up to 100x100x50 cm*

If achieving the carry size can happen only in folded mode, then the folding function should be described or at least allowed in the requirements.

Here comes one important point - specifying the folding function, assuming that it will have wheels, etc. starts to dilute the solution-neutrality of the requirements. This is normal - it is up to the wisdom of the systems engineer to constrain the design decisions only where needed, and to leave them free where the specialists will know better.

What we have seen here are, conditionally speaking, two levels of requirements - "user" and "system" requirements. The *user* requirements are less strict and often non-technical, presumably unfiltered input from customer or user research. The *system* requirements are more specific and respect the rules of "good requirements" much more. The system requirements will set more strict or quantifiable targets, like "...weight of 850g" instead of "...shouldn't weigh a lot." It is up to the judgement of the systems engineer to determine which user requirements to translate to system requirements and which to leave unchanged. The output from the requirements engineering step may look like Figure 17.

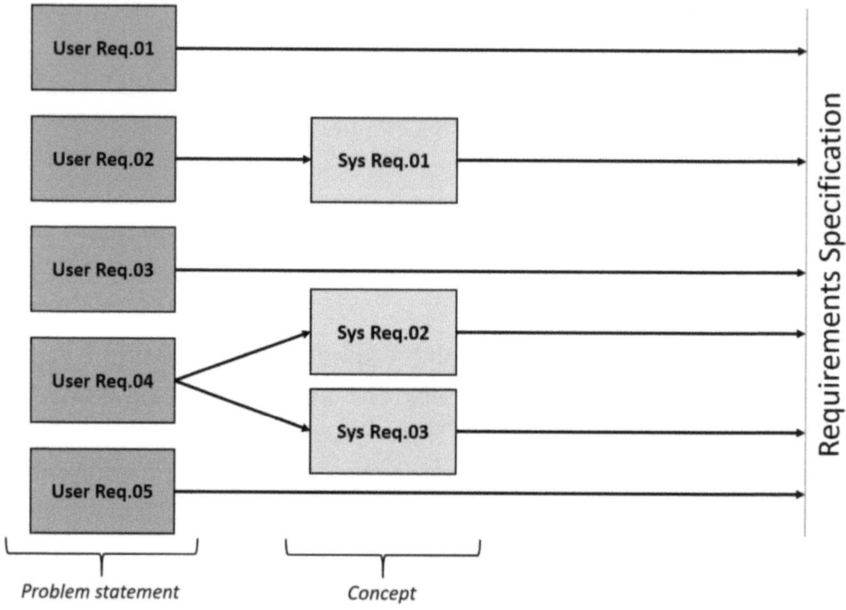

Figure 17 *User and system requirements in the specification*

> *The requirements may have different levels of details and abstraction and are not necessarily "good" or "bad" due to this.*

There is no hard rule as to how precisely the requirements specification should describe the system. Imagine you are reading a book where the protagonist walks through a forest. The portrayal of the forest can be simply "a forest," and every reader imagines the scene differently. If the book becomes a movie, great filmmakers may do a brilliant scene with the forest, given their creativity and talent and the book's freedom. On the other hand, the forest may be described in a whole page of spiritual narrative with such detail that if someone makes a movie based on the book, the forest will appear on the screen exactly as the readers have imagined it while reading. Do you see the range of options? If we want to come up with the most innovative product design, we should leave it as abstract as it could be (and constrain as little as possible the creativity of the people). If we already have some idea of what our product should be, we may need to step further in the design direction. Here is where requirements engineering fuses with *concept design*.

> *The requirements specification may state more than just problems...*

Let's discuss now which known solutions could fit our requirements specification from above. These (Figure 18) are some "usual suspects" - a scooter, bicycle and skateboard.

Figure 18 *Possible known solutions*

They may perform differently in many aspects, but each of them will satisfy our Vehicle requirements. The systems engineer should ask - do I really care if I get a skateboard or a bike? If yes, the systems engineer may steer the engineering in the direction of one of those vehicles, or even directly name the Vehicle a "Bicycle" for example.

Now, assume we still stick to our Vehicle requirements specification, but we don't want to get a scooter, bike, or skateboard. Prohibiting those designs in clear text will not lead to a nice specification, and even less to a good result. Here is where the concept ideation comes into play - the systems engineer may define a high-level conceptual solution and use it to further describe the system.

There are many different ways to generate ideas about the concept, from simple creative brainstorming to formal methods like Design Thinking. But here we will only illustrate the process using two simple tools: a ranking matrix and morphological analysis.
First, we may wish to rank the known solutions over a set of common criteria. The selection of the criteria should be done with consideration of the main objectives of the product and the project. Below is a sample ranking matrix for the three known solutions:

10 - best, 1 - worst	Scooter	Bicycle	Skateboard
Speed	6	10	4
Range	5	10	3
Size	7	4	10
Weight	8	3	10
Cost	7	4	8

From this table we can see that there is no clear winner, as each solution has its strengths and weaknesses:
- The bike is the best performing variant, but due to cost, size, and weight, it may not be the optimal one for our use case.

- The skateboard has the lowest performance as a commuting tool, but is very small, light, and affordable.
- The scooter seems to be a compromise between both and may be the winner for the application.

Instead of choosing between these existing options, we may try to design our own product using the strengths of the known solutions. Let's see how each solution implements the main functions[1] of our Vehicle (*propelling, steering, position*/support of the user):

	Scooter	Bicycle	Skateboard
Propelling	Kick-push	Pedal-driven	Kick-push
Steering	Handlebar & Stem	Handlebar & Stem	Trucks
Position	Standing-Handlebar	Seated-Handlebar	Standing-Free

The *propelling* type of the bicycle (pedal-driven) contributes to its high performance in speed and range (not entirely, but significantly). We may combine this with the factors contributing to the compact size of the scooter and the skate - the standing *position*. The *steering* via trucks is an additional contributor to the compactness of the skate, which we may also try to consider for our Vehicle. The morphological analysis can do exactly this - combining solutions for each of the mandatory functions to generate new designs. Reflecting our discussion in the matrix will show us this one possible alternative concept (as a result of the highlighted solutions):

---

[1] *More details of the functional analysis will be seen in the next chapter.*

	Scooter	Bicycle	Skateboard	New Concept - 1
Propelling	Kick-push	Pedal-driven	Kick-push	Pedal-driven
Steering	Handlebar & Stem	Handlebar & Stem	Trucks	Trucks
Position	Standing-Handlebar	Seating-Handlebar	Standing-Free	Standing-Handlebar

If we continue to combine the solutions, we may come up with even more concepts (Figure 19 shows them visually).

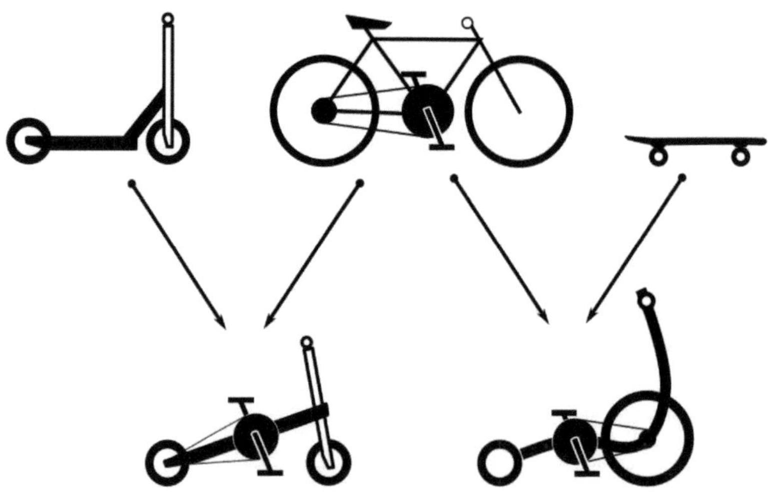

Figure 19 *Graphical view of the morphological analysis*

Well, if *Micro's Pedalflow™* (left image) and *Kolelinia's Halfbike™* (right image) hadn't been invented already, we could have invented them right now! :)

It is important during the ideation process to consider several solutions to the problem (alternative concepts). Needless to say, each concept should satisfy the mandatory requirements - a solution *not* to our problem is *not* our solution, no matter how good it is! There should be also an objective criteria to select the "best" solution to continue with - usually elected with a decision matrix or similar decision-making tool. We may even use the ranking matrix from above (of course, this is very simplified example using non-weighted criteria, etc.):

10 - best, 1 - worst	Scooter	Bicycle	Skateboard	New Concept - 1
Speed	6	10	4	9
Range	5	10	3	9
Size	7	4	10	6
Weight	8	3	10	7
Cost	7	4	8	6
RANKING	**33**	**31**	**35**	**37**

It's better to describe a concept in the specification than to artificially avoid the obvious.

A requirements specification based on a concept design is actually very common in practice. It is more directive and steers the engineering to achieve the desired solution. Some requirements may not be "very" solution-neutral, like:

- *Use wheels with inflatable tires*
- *Use pedal-driven propelling*
- *Shall utilize skate-like trucks for steering*
- *Shall have a foldable frame*

We may even include charts and sketches in the requirements specification to minimize the risk of interpretation error.

I have put the word "very" above in quotes because even the solution neutrality is relative. From our Vehicle perspective, the wheel is already directed by the specification, but for the designer of the wheel, this requirement is fully solution-neutral: materials, bearing, hub, spokes, dimensions, etc. are not defined.

> *The context of the system is not only the static environment. The time and the lifespan are context too...*

So far, we have been discussing the context of the system (Figure 16). We shouldn't forget, though, that the context is not only a static view of the environment where our system is supposed to serve, but also the time span of the system, from the moment it is built until its retirement and scrap.

Let's look at two wristwatch products: a mechanical watch and a smartwatch. While they may appear to belong to the same product group, these are actually two very different products:

- Have you seen the wristwatch of your grandfather? It may be still functional after many decades of service. The mechanical watch is a kind of product which could have a lifespan of many years. When designing such a product, it is important to consider long minimum lifespan requirements for the materials and construction, and a high level of serviceability and parts availability.
- If you are one of the early smartwatch users, you have probably already thrown away your first smartwatch. The obsolescence of the smartwatch comes from many factors - some physical (aging of the electronics, battery, display), some social (new trends, new models, new features), and some commercial (software deprecation, incompatibility with the up-to-date information ecosystem). Usually, a smartwatch has an active service life of only a couple years. It isn't a product you usually repair. Instead, the design and materials should allow easy recycling, biodegradability, and a low environmental footprint - define those as requirements!

> *The system definition should consider the requirements for service, reliability, longevity, sustainability, etc.*

The next chapter will address the architecting activity. It looks like it follows requirements engineering (Figure 12), but in reality they partially overlap and interact with each other (Figure 12A), as the design decisions may generate additional product requirements (for manufacturing, suppliers, and different disciplines). In addition, decisions on trade-offs in the design may lead to revision of the initial requirements as well, in order to reflect the new system solution. Last, but not least, after the system architecture is complete, we may generate entirely new requirement specifications for some of the subsystems or parts. One very typical example is the car. The car manufacturer (the so-called OEM - think of Daimler, BMW, Nissan, Toyota, Ford...) creates the concept and architecture of the vehicle, makes a system decomposition, identifies the needed parts, and has a clear idea of what is required for each part and which interfaces will connect it to the rest of the system. When a supplier gets the requirements specification for one of those parts, very little from the product concept is left to the imagination. The implementation details are usually quite loose, but sometimes the customer may direct the CPU, network transceiver ICs, or operating system - based on strategy, technology, or supply-chain decisions.

> *Everything is relative, including "good requirements" rules too...*

As final words to this chapter - my practical advice is don't blindly strive to satisfy the definition of "good requirements." The level of abstraction is very dependent on the context and shouldn't be perceived as an ultimate goal. Often it makes no sense; for example, consider when designing a spoon (which everyone knows), and the requirements try to describe it as "a way to transport food of various consistencies to the mouth."

Sometimes excessive effort to reach atomicity leads to human-unreadable textual monsters - don't forget that your colleagues are going to work with them every day (we know the requirement specification isn't poetry, but it shouldn't be a struggle to read either). The whole set of requirements engineering activities shown here is, of course, not always needed - if you are an established bike manufacturer, you will never start with a "point A to B" problem statement, but directly with the detailed bicycle system specification, knowing your target, your current model portfolio, platform, and component base.

Applying common sense is as always mandatory and requirement engineering makes no exception.

Design the Solution (System Architecture)

We have already touched on the architecting activity in the previous chapter - remember the concept ideation part? In architecting, we should go a step further and design an actual physical *solution* to the problems stated in the requirements specification. The system architecture is the description of this solution.

Sometimes the requirements are not enough input to start working on the physical architecture, and we may want to make a functional architecture (functional analysis) of our solution first. This analysis aims to define the set of functions the system should have, in order to be able to resolve the problem. Some of the functions of complex systems are called *emergence*. Emergence is a behavior which is a property of the system only, but not of any of its parts separately. Emergence, being a result of complex interactions, is sometimes non-deterministic and doesn't always follow the obvious cause-effect paradigm. Emergence in a system could also be hard to predict based only on knowledge about its parts. Consider the *transportation* function, from the example given in the "Introduction to Systems" chapter, as the *emergence* of our Vehicle system there. It is important to know this terminology, but not necessary to use it as a classification of a function or behavior for now.

> *Emergence is a property of the system which none of the parts have alone.*

In the example with our Vehicle from the "Requirements" chapter, we actually did one functional analysis - we have defined the following functions our solution should have: *propelling*, *steering* and *position* (not exactly a function yet, but we will make one of it). Now that we know about the functional analysis, let's define more thoroughly the functions we would like for our system:
- *Slide on the road*
- *Propelling*
- *Steering*
- *Contains the user (in standing position)*
- *Braking*
- *Ensure structural integrity*
- *Enable folding*
- *Insulate the vehicle from shocks from the road*

How long this list will go depends on the level of detail selected by the systems engineer for the architecture. In more complex systems, the functional architecture is not even a list, but can have a hierarchical or graph structure, relations between the functions, etc. To ensure that the functions are there to serve the needs, a traceability should be established between the functions and the requirements.

We have already established a traceability once in the previous chapter (look at Figure 17). We will do it again now with one requirement as an example:
- *The Vehicle shall be compact and able to be carried on the public transportation*

The functions which directly serve this are:
- *Ensure structural integrity*
- *Enable folding*

Although neither of the functions are enough to entirely satisfy the requirement, they can enable it in this case. If you remember from the "Requirements" chapter, we translated this user requirement to two system requirements:
- *Weight up to 18 kg*
- *Carry size up to 100x100x50 cm*

These should be considered as constraints to the physical solution.

We may want to design a functional solution first, before we make a physical one.

After the functional analysis is done, we can move on with the physical architecture. The functional architecture adds even more inputs for the physical solution - a set of user- and systems-requirements (as shown on Figure 17), and the functional model of the solution in addition (Figure 20).

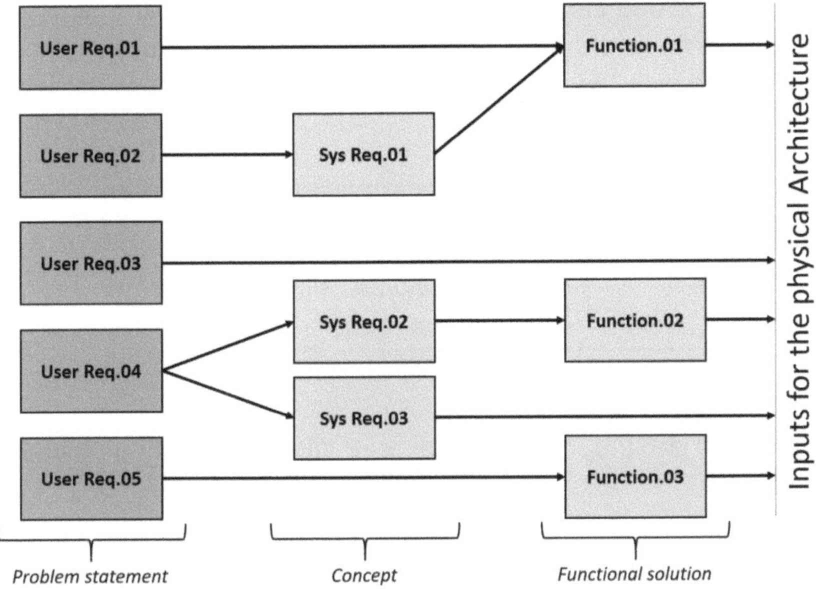

Figure 20 *Functional decomposition as input for the architecture*

Finally, the system architecture shall describe the physical system, which is a solution to our initial problem. As we have seen in its definition, a "system" is a group of parts and interactions between them, which we usually call "components" and "interfaces" in engineered systems. This means that the physical architecture shall at minimum identify and represent this group:

- All parts / components
- All interactions / interfaces
- All system-relevant critical design and performance characteristics of the components and the interfaces, as well as their expected behavior

This concerns the system **inside** our system's boundaries. But to model the solution adequately, the important external interactions with the system's context should be considered in our architecture as well - like the user, infrastructure, or time - anything that might be applicable. Good selection criteria may be simply to ask "What will happen to my system if I add/remove [context element]?" and be very critical with the answer, if we don't want to include the whole world as a relevant context :)

The architecture is a description of the physical solution to a need.

In addition to the components and interfaces, the architecture should:

- Assign each function from the functional solution to the physical subsystem which delivers it (a set of components and interfaces). The same should be done for the requirements which "reach" the architectural solution as inputs (from the Figure 20 above). The ultimate purpose of our solution is to resolve the initial problem and this shall demonstrate it.*
- Decompose the system to subsystems. In multidisciplinary systems, the subsystems may be software, electrics/electronics and mechanics, for example. In larger-scale systems, like airplanes, each discipline may have many subsystems, sub-subsystems, and so on. You can imagine the complexity of an airplane - even its mechanical construction needs to be decomposed into multiple mechanical subsystems and sub-subsystems, and the same is true for its electronics and other subsystems.
- Show the construction / structural solution.
- Demonstrate the dynamic behaviour of the system and link it with the system's use cases. This may include its start-up

sequence and timing, shut-down sequence, user interaction, processing, etc.

- Communicate design directions for the components and interfaces, if the implementation or the function is relevant to the system. This is not a rule, but the interfaces are more likely to have this property.
- Define the verification criteria for the correct implementation of both components and interfaces, according to their system relevance.

* Seen from the side of the physical system (i.e., which physical subsystem/component is relevant for which function/requirement), this assignment could also indicate to us how optimal the solution is, related to the problem. Do you remember the trap designs from old cartoons - a very long chain of actions between the trigger and the trap (boots kicking balls, candles burning ropes, weights falling on a balanced plank, etc.)? All these components and interfaces are not doing anything except making the system ridiculously complex (and as a result - funny). The functions of the automatic trap are usually assigned only to the beginning (the trigger mechanism) and the end (the trap itself) of this chain, and everything in the middle is not relevant for the purpose of the system and serves only its "design decisions."

> *The architecture may be documented in many ways - formal or not.*

Now, in the spirit of the book, it is time for an example. There are many ways to document the architecture and there is no absolutely wrong one - documentation can take the form of written documents, text and drawings, formal or informal diagrams, formal models using UML, SysML or OPM, or a simulation in Simulink. This book is dedicated to introducing one minimalistic tool for systems engineering, the I-CM (Interface-Component Model) and we will see later how we can even use a spreadsheet to work with the system model. But before we reach this (very easy, yet formal) method, let us do it in a more natural way for now (Figure 21).

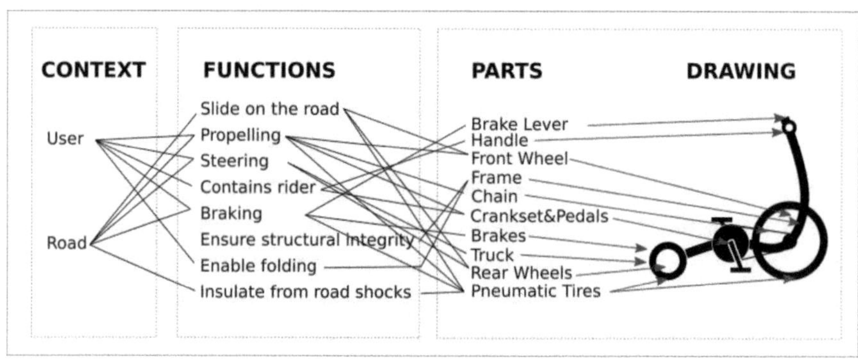

Figure 21 *Form-to-function mapping - graphical view*

What we have in this view (in the dotted boxes) is the physical context (Context), the functional view (Functions) and the physical view of our solution (Parts and Drawing). Linking the functions and parts is sometimes called "mapping function to form." The physical context is related to the physical view of the system, as both are from the physical world, but one belongs *out* of the system boundaries, and the other one *inside* the system boundaries. Again, only by considering the relevant context can we adequately design our solution: for example, when the user is a factor, we can design a solution with the user's weight and size, physical strength, and anatomy in mind; i.e., make a solution for a real human being. We can generate constraints as well, from the system to the context - for example, that the Vehicle is suitable for users only above 40 kg and less than 100 kg.

If we continue adding more relations to the graph representation from Figure 21, it will soon become unreadable, so we need to divide it into several views and use text and maybe even tables to make the architecture document manageable. An easy and common way to do this is to put the "form" and the "functions" as two dimensions in a matrix and mark the intersection of relations in some way (Figure 22).

> *Establishing a link between the form and the function is essential for the architecture.*

	Brake Lever	Handlebar	Front Wheel	Frame	Chain	Crankset& Pedals	Brakes	Truck	Rear Wheels (x2)	Pneumatic Tires	EXT: User	EXT: Road
Slide on the road			X						X	X		X
Propelling			X		X	X				X	X	X
Steering								X	X	X	X	X
Contains rider		X				X					X	
Braking	X						X		X	X	X	X
Ensure structural integrity				X								
Enable folding				X							X	
Insulate from road shocks										X		X

Figure 22 *Form-to-function mapping*

Yet something very important is missing from both Figure 21 and Figure 22! We see that most functions are linked to more than one physical part (component), which means those components must interact to deliver the function. That's right - the **interfaces/interactions** are missing, and if we refer again to the chapter "Introduction to Systems," we know that the system functions (emergence) are impossible without the interactions of the parts. The interactions are not only between the parts inside the system boundaries, but also to the relevant parts of our context, like:
- The *User* stays on the *pedals* and holds the *handle* to be contained on the vehicle.
- The *User* rotates the *pedals* to propel the vehicle.
- The *User* pushes the *brake lever* to decrease the speed of the vehicle.
- The *User* shifts his weight to steer the vehicle via the *truck*.

> *With the I-CM, we can model the system by identifying its parts and their interactions.*

Different systems engineering tools and formalisms have different ways to represent the interfaces. In this book, I introduce my method, the I-CM, which we will apply now (Figure 23). The idea is to represent the physical system by listing the components on one dimension of a matrix, and the interfaces on the other. The interfaces here are all the intended[2] interactions between the parts of our system (like the bearing between the wheel and the frame, which constrains the movement in one direction only, etc.). In the intersection between an interface and each of the components it connects, we mark a relation. In I-CM, the relations are of several different types, but for this first I-CM example, let us use simply "X" to indicate that there is some relation.

	Brake Lever	Handlebar	Front Wheel	Frame	Chain	Crankset& Pedals	Brakes	Truck	Rear Wheels (x2)	Pneumatic Tires	EXT: User	EXT: Road
Bearing front			X	X								
Bearing rear (x2)								X	X			
Friction to the rims							X		X			
Bearing - central				X		X						
Bead press			X						X	X		
Power transfer			X		X	X						
Braking signal	X						X				X	
Structural fastening		X		X								
Structural fastening	X	X										
Pivot connection				X				X				
Support position		X									X	
Apply force						X					X	
Carry						X					X	
Contact with ground										X		X

Figure 23 *Interface-Component Model (I-CM), defining the interactions between the parts*

---

[2] In further design analysis, the unintended interfaces should also be considered.

As we can see, the tables on Figure 22 and Figure 23 have the same columns, both containing the list of the system components. This allows us to easily observe many relations within the system if we simply adjust the two matrices. Let's look at the *braking* function in the matrix from Figure 22. It is delivered by the user, brake lever, brakes, rear wheels, tires, and road. From the table in Figure 23, we can find the interactions between those components and see the whole Braking subsystem: user (braking signal), brake lever (braking signal), brakes (friction to rims), rear wheels (bead press), tires (contact with ground), road.

As with concept ideation, alternative architectures should be considered during the architecting process. The alternative solutions could start with different functional architectures, but also the same functional solution may generate alternative physical architectures. The final architecture should be the result of selecting the optimal solution from all alternatives. We can hardly use the word "best" instead of "optimal," as almost every solution is a compromise between several trade-offs. During the architecting, it is important to identify the most significant trade-offs and properly document the decisions. In our example, we may have trade-offs like:
- Weight vs. speed
- Cost vs. weight
- Weight vs. robustness
- Size vs. range

These examples are intentionally simple and limited to only two counter-related parameters. In complex systems, there could be multiple parameters and non-linear relations. Deciding between trade-offs is usually done with thinking and expertise, and not applying a decision pattern. Some methods, however, try to formalize this process (like TRIZ for non-deterministic systems and Linear Programming for mathematically-modelled systems).

Designing a solution always includes resolving trade-offs.

For better understanding, let's see some typical examples of trade-off decisions:

- Manufacturability vs. Maintainability: Manufacturing usually hates screws. Replacing screws with snap-fits is very much preferred when it comes to fastening two or more parts. But when it comes to a product whose life expectancy is very high and maintainability is a critical requirement, screws are a much better option for multiple assembly/disassembly cycles than snap-fits. *Additional note:* The architect should closely cooperate with the manufacturing team. The design we create should be manufacturable (if it is a service or software, it has similar requirements but for deployment, accessibility, etc.). Usually review meetings are required for both parties to consider the constraints of the other side and come to a consensus.

- Software vs. Hardware: In embedded audio systems, specialized DSP (Digital Signal Processing) chips are usually preferred to process the audio (for equalizing, filtering specific frequencies, etc.). A software DSP implementation on a powerful enough microcontroller may be a more cost-effective solution than having an additional DSP chip and slower microcontroller. Or vice versa - but in any case, it is a typical trade-off decision made by resource estimations, sound quality requirements, and cost comparison. *Additional note:* A typical discipline decomposition is seen in this example - what goes to electronics design (EE engineers) or to software design (SW engineers), how they will interact, and where are the boundaries of the respective subsystems.

- Solution vs. Solution: Consider aluminium housing (assembled with screws) used to cool down electronic components while acting as a radiator vs. plastic housing (snap-fit construction) and active fan cooling. If BOM and assembly costs are the same, then some additional functions or properties should be analyzed, like: Is the metal housing going to have additional functions as well (like EM shielding, which plastic can't do)? Is the lifetime of the fan critical (being an electromechanical component)? Is the noise of the fan acceptable for the application? Is the cooling performance comparable in different environments? Is the weight difference considerable? *Additional note:* This example pushes multi-factor trade-off analysis, including contextual factors like life expectancy, environmental performance (electromagnetic compliance and thermal), and manufacturing complexity.

- Make vs. Buy: This may be applicable for every subsystem. Technical risk, know-how, cost, maintainability, dependency of a third party, prospective re-use of own future subsystems (platform strategy) - these are some of the many factors which may be considered in making such a decision. *Additional note:* In this example, we have two aspects - platform strategy (re-use of existing development, or developing something with a strategy of future re-use) and supplier management. When we decide to go with a supplier (the Buy option), the respective subsystem should be described in a detailed requirements specification if we want to receive the subsystem we need instead of what the supplier imagines.

- Safety vs. Cost: In the chapter on the requirements engineering, we mentioned that the requirements specification should not be redundant. In the design, redundancy increases the complexity and cost. But this doesn't necessarily mean that redundancy should be universally avoided! In safety-critical systems, redundancy increases safety. A few examples are redundant communication channels, computation, data storage, or power supply. *Additional note:* For example, showing critical information to the user may be organized by computing it twice in different subsystems. Information could be shown only if the two sources give the same result; otherwise, an error notification or safety reaction could be triggered. This kind of redundancy is a desired mechanism, chosen due to criticality of the safety function, against its increased complexity or cost.

- Adaptivity vs. Complexity: Some systems use feedback loops in their operation, which allows them to adapt their performance to changes in the environment, other context, or other parts of the system. Other systems, however, simply perform without taking into account how their performance manifests in the real world. Adding feedback loops (including from self-diagnostics, measurements, monitoring, etc.) will increase the complexity of the system, but may significantly influence the system's performance. As an example, think of an audio system for an environment with variable noise (like in a vehicle). On volume setting X, a non-feedback system will play the music always with a certain volume, independent of the environmental noise, and the passenger will perceive the music as being quieter when the car is going faster or with open windows. A system with feedback will apply the output volume setting, taking the background noise level (monitored) as a "zero point" and adding the output setting on top of it. In this way, the absolute output of the audio system will increase as the background noise increases, but the passengers' perception of the music volume will be constant. *Additional note:* Need of feedback in the system may come both from specifications and from the solution. Systems with feedback sometimes have completely different performance, and sometimes the feedback adds little to nothing. This isn't a trade-off with a default solution. While the example with the adaptive volume setting may be a good solution for an automotive infotainment system, for home audio it is unnecessary complication.

- Design vs. Manufacturability: Designing a product which can be easily manufactured and whose design prevents manufacturing mistakes sometimes requires compromises compared to the construction of a single object. Using poka-yoke (preventing wrong assembly), unified components (for example, having all screws be of the same type), or minimizing handling (top-to-bottom assembly steps) may seem like unnecessary constraints, which may lead to compromise with the aesthetics of the product. However, they can ensure consistent high quality of a serial product of thousands or millions of copies. *Additional note:* As in the first example above, feedback and suggestions from manufacturers and suppliers are useful input during the design and it may require several loops to reach a good balance between trade-offs.

> *Think of how your system may fail, and try to resolve it.*

Another valuable activity during the architecting process is to perform a failure mode analysis. We can take every function from our functional solution and analyze:
- How can it fail?
- What will be the impact in case of failure?
- What can we do to prevent the failure from happening?
- How can we minimize the impact if the failure happens anyway?

Do the same failure mode analysis with the components and the interfaces. It sounds like a lot of work, but it will pay off, and in addition - many failure modes will have the same preventive measures as they may trace to the same potential root cause. For example - one of the reasons the *propelling* function of our human-powered Vehicle could fail is from a failure of the drivetrain. The chain component may fail if it breaks due to overload or tear. The proper design and sizing of the chain is a countermeasure for both entries. Most of the *preventive* measures for failure modes should be propagated as design decisions to either components or interfaces. The *detection* of the failure mode during the product qualification (testing) should be added to the verification criteria of the components, interfaces or functions - depending on the failure mode.

One type of failure of physical components is (pre-) damage during the transport or manufacturing. The potential failure modes due to the manufacturing process should be communicated to the manufacturer so that they are considered with special attention in the manufacturing plan. Similarly, the transportation damage modes should be communicated to and considered by the logistics and packaging teams, and so on. For more details about this, refer to the "FMEA" in the external literature.

In the next chapter, we will learn about system integration. For this activity, a system architecture is the most important logical input, as it is the guideline on how to assemble and verify the physical system. Some important insights on how to do a system architecture will be discussed, so please read further.

Integrate the Parts (System Integration)

If we refer back to the V-model in Figure 12A, system integration is the first activity which appears at the right side of the "V." It has the *ready parts* and *subsystems* as physical input, and the guidance of the system architecture on how to assemble them into a finished working system. The two words here - "assemble" and "working" - actually define the two main objectives of the integration process - we not only put all together, but we also verify that everything works together in the "right way." The right way still doesn't mean that the system resolves the initial problem. It means that the system works as intended per the design.

> *System integration is the activity where the physical parts come together to build the system.*

Let's start with assembling the system. This is *not* the assembly process on the production line, although some relation exists. During system integration, the parts and subsystems (as per the architecture) are put physically together, connecting them via their interfaces. Establishing the interfaces might be the biggest part of the integration - this includes connecting wires; screwing, welding, and glueing parts; connecting power supplies; and configuring data protocols, communication channels, software-hardware bring up, etc. The integration process may be done in two general ways:
-   Big-bang integration: All the parts are put together at once. There's a high risk of it not working and having hard-to-identify issues, but it requires less preparation and planning. This is usually acceptable for simple systems.

- Incremental integration: The parts are added incrementally to the build, always achieving a stable intermediate assembly. The sequence of the integration should be carefully planned, considering all the dependencies and prerequisites of each part or subsystem.

For us, the incremental integration is more interesting, as it is reliable and suitable for systems of any scale. The analysis which can be performed with the I-CM tool (see the chapter "Introduction to I-CM") may be used to plan the integration steps. The system should be partitioned into relatively autonomous clusters, which once assembled could be tested for integrity and functionality. If a cluster has unresolvable dependent interfaces to components outside of itself, an instrumented environment may be created to support the integration of the cluster. Often, the components themselves are developed within a simulated environment (the components from this simulated environment are called "stubs" in software engineering) to enable the component engineering to finish without the whole system available. Using part of this simulated environment during the incremental integration can allow controlled integration without involving too many additional components due to dependencies. Figure 24 shows a graphical representation of this approach.

1. Previous integration step (in place of Component D, there is a stub providing the interfaces needed from Component A and Component C):

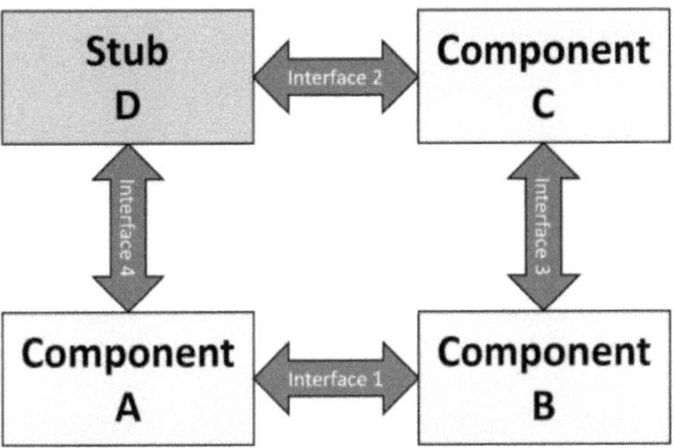

2. The new component (Component D and its stub development environment), to be integrated in the current integration step:

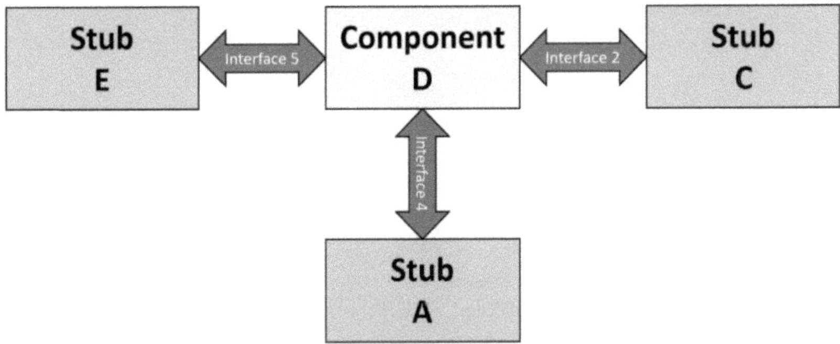

3. Current integration step (adds Component D and temporarily the stub of Component E to allow the assembly to function):

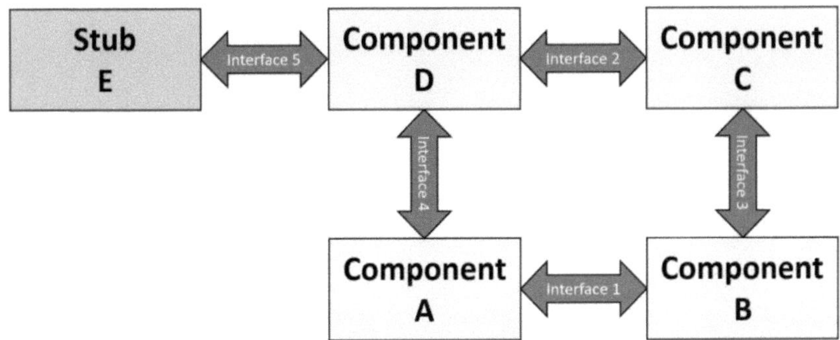

Figure 24 *Incremental integration*

Let's assume that we are a car manufacturer and our future Vehicle has the following E/E architecture (Figure 25).

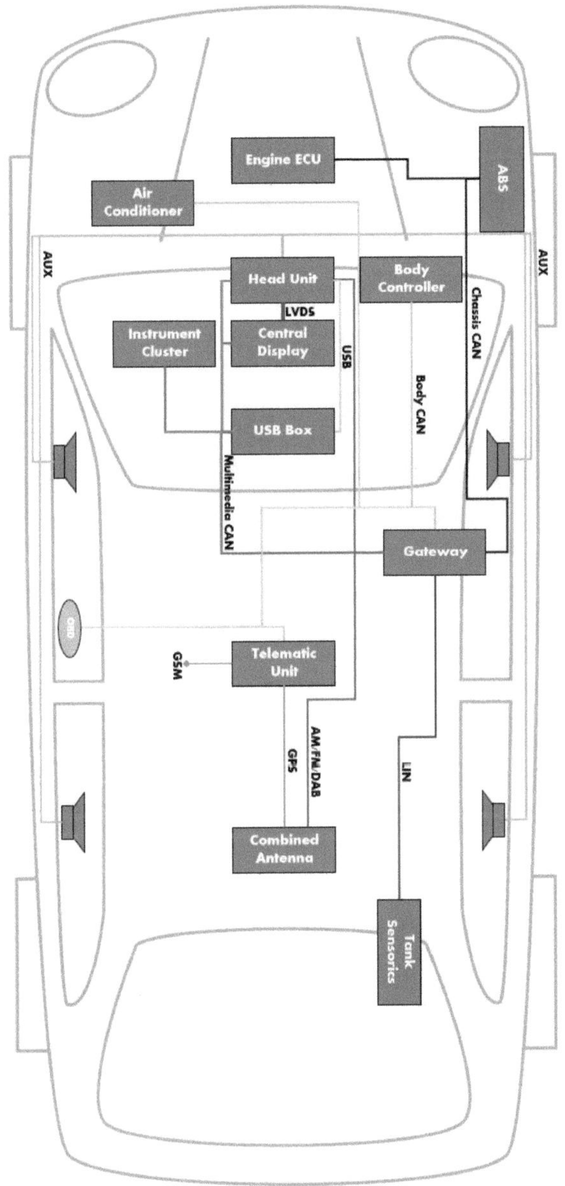

Figure 25 *Automotive E/E architecture*

What we can observe here is that there are four network domains, serving different subsystems of the vehicle - Chassis CAN, Body CAN, Multimedia CAN, and LIN. We may plan our integration steps by introducing the devices from one network at a time, for example. Still, there will be many dependencies which may prevent the partially-integrated system from running and being tested (Figure 26 shows the I-CM of the vehicle E/E architecture).

Take, for example, the feature *display fuel level* of the instrument cluster. We will have logical dependency on the information coming from the tank sensor. The physical dependence is the interface between the instrument cluster and the tank sensor, which in this case includes the gateway as both functionally dependent units are in different network domains - the information from the tank sensorics is broadcasted in a LIN network, while the instrument cluster is connected to the Multimedia CAN.

Components → / Interfaces ↓

Interfaces \ Components	uC (microcontroller)	RTC (real-time clock)	LCD (display module)	Body Controller	Head Unit	Central Display	Speakers	USB Box	Instrument Cluster	Gateway	Microphone	Telematic Unit	OBD	Combined Antenna	Tank Sensors	EXT: GSM Network	EXT: OEM Back-end		Interface connectivity
I2C Network for RTC and LCD (uC Master)	IO	IO	IO							IO									4
Body CAN			IO	IO						IO		IO	IO						5
Multimedia CAN					IO	IO		IO	IO	IO									5
LIN										IO					IO				2
LVDS					IO	IO													2
USB					IO			IO											2
AUX					O		I												2
MIC analong signal					I						O								2
AM/FM/DAB antenna signal					I									O					2
GPS antenna signal												I		O					2
GSM/LTE connection (Infrastructure)												IO				IO			2
Data gateway to mobile providers																IO	IO		2
																			0
Component connectivity:	1	1	2	1	6	2	1	2	1	4	1	3	1	2	1	2	1	0	

FUNCTION / REQUIREMENT	uC	RTC	LCD	Body Controller	Head Unit	Central Display	Speakers	USB Box	Instrument Cluster	Gateway	Microphone	Telematic Unit	OBD	Combined Antenna	Tank Sensors	EXT: GSM Network	EXT: OEM Back-end		DIST
									Component Interactions										
IC: Display Vehicle Speed	1								1	1									3
IC: Display Motor RPM		1							1	1									3
IC: Display Blinker Status				1					1	1									3
IC: Display Lights Status				1					1	1									3
IC: Display Fuel Level									3	2					1				3

Figure 26 *I-CM of the automotive E/E model*

But let's refer to the incremental integration with the use of stubs (Figure 24) and we may easily find a way to build our testable subsystem. We don't need to add all the physical interfaces beyond the boundaries of our "current integration step" subsystem; it is enough to introduce a network simulation of the rest of the E/E system (Figure 27). The network simulation (a computer program) does not even need to simulate the whole functionality of the components, but only their network behaviour. The tank sensorics simulation does not need a gateway in order to send the message directly to the Multimedia CAN. The telematic unit simulation does not need the physical GPS signal from the antenna to broadcast GPS coordinates in the network, etc.

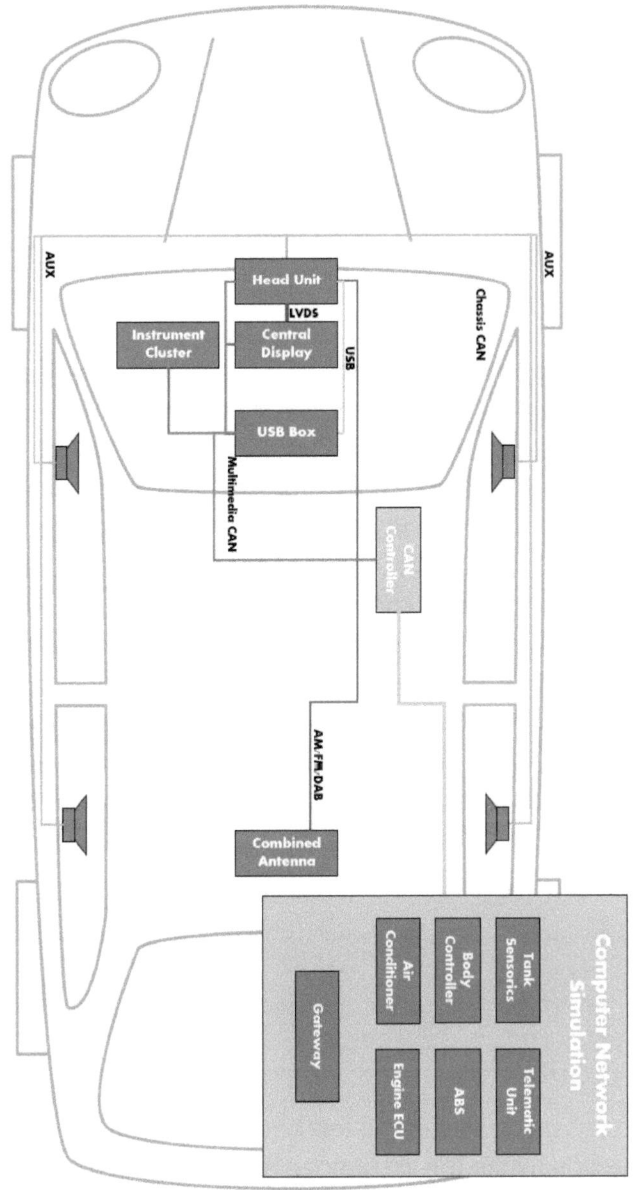

Figure 27 *Incremental integration of the automotive E/E system*

> *During the integration, the system is still an open box - we can observe how all the "internals" work together.*

Assume we finally have an integrated system. Before we test it for resolving our initial problem (doing the right things), we need to make sure that it works as intended per design (doing the things right). The system we have for testing during the integration is a white-box (or open-box) system. We can, and should, test its internals - the interfaces, timings, performance - all those things not visible once the system is finished and the "box is closed." Testing the interactions between the parts (interfaces) is one of the most significant activities here. Remember that the specification of the interfaces and the ways to verify them (once implemented in the system) are described in the system architecture. In this chapter, we will see some examples of interfaces from different disciplines and how the expected interface details could be addressed in the system architecture.

> *And the interfaces are mysterious again...*

There is a certain duality of the interfaces. We said that an interface is an interaction, while the components are more materialistic. Then, if we have a system of two devices, communicating over I2C with each other, is the harness between them the interface? Or does the system consist of three components (one of them being the harness itself) and the electrical signal is a function of the "harness" component? Purely theoretically - the second one is correct and the interface is the signal itself (voltage levels, form, etc.). However, sometimes the cable is considered the interface, and the signal is its property. In the chapter "Special Attention on Interfaces," we have seen that an interface may propagate a subsystem downstream in the design. Following this explanation and depending on the context, the interface may have a "physical" ingredient or not.

Now follows a few examples of interdisciplinary or system-relevant interfaces and how to treat them in the system processes. Of course, interfaces of that "level" will not pop up in the system architecture of large systems (like cars or airplanes - these will look more like what we have seen in Figures 25/26), but they may have relevance in a small system, like a home stereo.

Often the metal housing of electronic devices is used as a "shielding." This plays a positive role for the ESD (electrostatic discharge) and EMC (electromagnetic compatibility) performance of the product. Usually, the electrical ground (-) of the PCB is rooted to pads which have screw holes, then a screw connects the PCB to the housing, creating in this way a structural and electrical interface. Figure 28 illustrates this assembly.

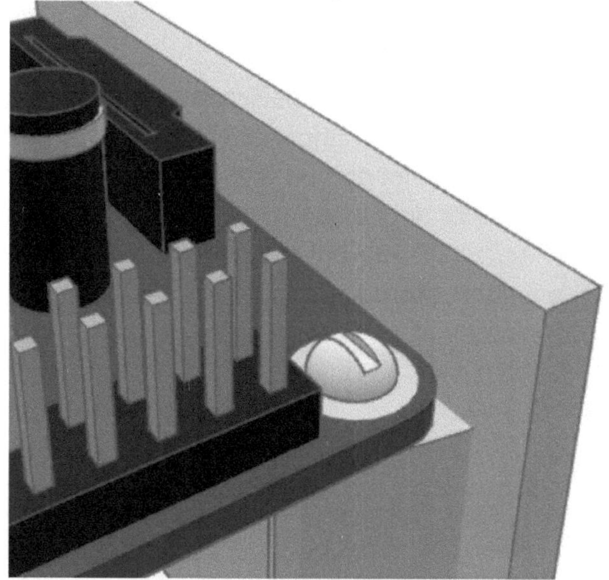

Figure 28 *Multi-purpose, multi-components interface*

Using the I-CM, we may represent this subsystem in several ways, depending on the targeted level of detail, the maturity of the engineering team, and other factors the systems engineer may consider. The most high-level representation may look like this and may be sufficient for a highly-experienced engineering team (Figure 28A).

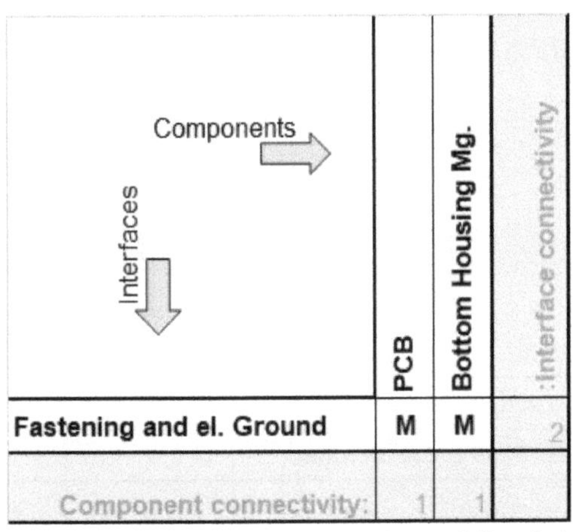

Figure 28 *High-level presentation of the interface*

A more detailed component and interface representation may consider decomposing the interface to its two separate functions and describing them in detail. We may consider the electrical flow as directional and serving the *grounding* function in a logical (and not necessarily physical) way - the energy comes from the PCB and goes to the housing, using the screw as a medium. The screw could be part of the representation, or not (and could be propagated by the mechanical interface in the detailed design). If we chose to have the screw as a component, the next figure shows the model with higher level of detail (Figure 28B).

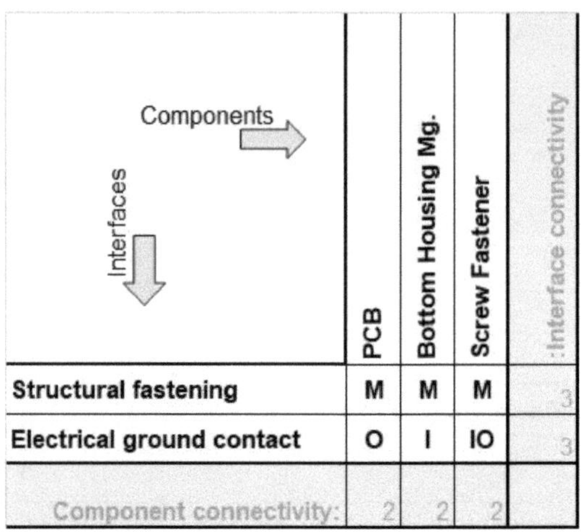

	PCB	Bottom Housing Mg.	Screw Fastener	Interface connectivity
Structural fastening	M	M	M	3
Electrical ground contact	O	I	IO	3
Component connectivity:	2	2	2	

Figure 28 *Detailed presentation of the interface*

In any case, it would be recommended to use the "design" and "verification" properties of the interfaces to describe some important design recommendations.

Let's take the "electrical" aspect of the interface and see what these could be:

DESIGN:
- Consider electrical resistance of the contact surfaces close to 0Ω over the lifespan of the product
- Select the materials of the involved parts to be compatible against galvanic corrosion

VERIFICATION:
- Perform ESD testing on the housing and accessible EE parts (ports, slots, etc.)
- Perform comparative EMC testing with and without being grounded to the housing, to evaluate the effect the interface brings

- Perform lifetime tests under high-stress climatic conditions, and inspect the contact points at the end of the test - measure the electrical paths and perform an optical inspection

A little background information - if we have magnesium housing, nickel-plated screws, and OSP finish on the PCB, the interface may not necessarily work consistently over time. The OSP is a coating which ensures no oxidation on the copper pads of the PCB until the assembly process is complete (say, 6- to 9-month shelf life). It degrades over time, which leaves the copper exposed to air, leading to an oxidation film with potentially high electrical resistance. This may lead to loss of electrical contact to the housing as the product ages. In addition, preliminary analysis for galvanic corrosion between nickel-magnesium, nickel-copper and magnesium-copper should be performed to confirm the initial design choice of the materials. The tests should confirm the analysis.

The "mechanical" aspect of the interface seems trivial, but some considerations may be:

DESIGN:
- The screwing process should be calibrated not to crack the PCB or the screwing dome of the housing
- The geometry and material of the screw should allow it to self-tap into the housing
- The tolerance stack-up of all parts should fit their interrelations (the smallest PCB hole should fit the biggest screw shank, the biggest PCB hole should be smaller than the smallest screw head, etc.)

VERIFICATION:
- Perform vibration tests of the part (5g at 4-40Hz in all three axes for one hour) and inspect the part integrity after the test
- Disassemble the parts and observe the surfaces for wear and loss of material
- In case of released particles due to wear, assess the risk of electromigration or causing a short (are the particles electrical conductors?)

Why is it a system interface? Because it is shared between EE and ME.
Should it be addressed in the system architecture? Yes (depending on the scale of the system). Alternatively, it could be covered by the mechanical skill and propagated to electrical via failure modes in the FMEA.

end of the example

The interfaces may have a dynamic nature or lifecycle, and this should be considered as well during the integration testing, and prior to that, in the system architecture as interface specifications. We have already seen a potentially "aging" interface in the previous example. If components or interfaces are commonly used for projects, it may be a good idea to propagate those to design rules instead of describing them in each system architecture. A good example is:

Double-Sided PCB Soldering

Most electronic applications nowadays use SMD (Surface-Mount Device) technology. The technology of surface mounting is very effective:

1. A solder paste is printed on the solder pads of the PCB via steel stencil.
2. The components are applied on the board with a pick-and-place machine so they stick to the paste.
3. The board and the components are heated in a reflow "oven" until the paste is melted to bind the components to the pads.
4. After cooling down, all the placed components are soldered with solder joints to the pads on the PCB. Figure 29 shows one component soldered like this.

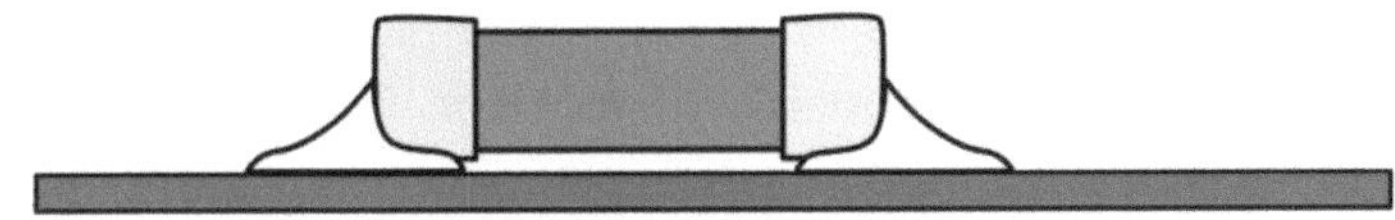

Figure 29 *SMD component - first reflow*

If the product requires a high level of integration (like for phones, laptops, tablets), the SMD components are mounted double-sided on the PCB. This means that the PCB should go through the above procedure twice, upside down the second time. But what could happen with the components that have already been soldered on the upper side during the reflow process with the PCB turned upside down? Usually, surface tension keeps the components in their places, even though the solder joints are melted again during the second reflow. But heavier parts may still get displaced or even fall completely off (Figure 30). Depending on the manufacturing practices (reflow parameters, solder paste type, etc.) a design rule may be propagated at the organization level to avoid placing heavier parts on the first reflow side of the board!

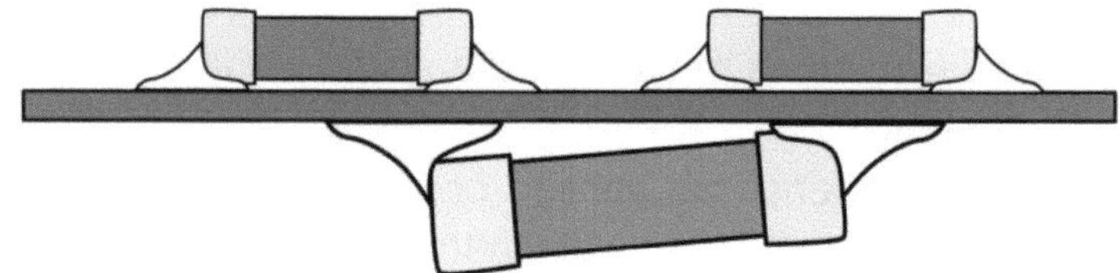

Figure 30 *Second reflow on the SMD line*

The solder joint is an interface between the component and the PCB with the help of the solder paste. This interface has different characteristics in the different phases of the life of the product. After the SMD process, the solder joint remains solid for quite a long time; however, this interface has a dynamic consistency during production. If we don't consider this factor, it may lead to defective parts. The dynamic of the interfaces should be seen from every possible perspective, including aging, its state in different operational environments (temperature, moisture, light), evolution of infrastructure, etc.

Why is it a system interface? Because it is shared between the EE/Layout and Manufacturing teams.
Should it be addressed in the system architecture? No. It is a better idea to add this as a design rule for the PCB-Layout team. It could be addressed from the Manufacturing team via the manufacturing FMEA to the Engineering team as well.

end of the example

The I2C network is a local two-wire communication protocol that allows several devices to share a single line and communicate with each other. The most common configuration is master-slave or master-multiple slaves. The master drives the clock/speed. Each device has its own unique address. The communication starts with START followed by an ADDRESS signal from the master, which every slave device in the network listens to. When the ADDRESS is recognized by a device, it receives the rest of the communication on its own (until a STOP signal). The protocol is used to transfer data or to command the devices. The communication goes in both directions between the master and any slave (I/O), but the slaves do not communicate with each other, as they can not initiate START and STOP signals.

In Figure 31, we see a part of the architecture of a digital desk clock. The microcontroller "uC" is the master of the I2C line, which connects it with the RTC (real-time clock) and LCD display module. The I-CM of this sub-system is shown in Figure 32.

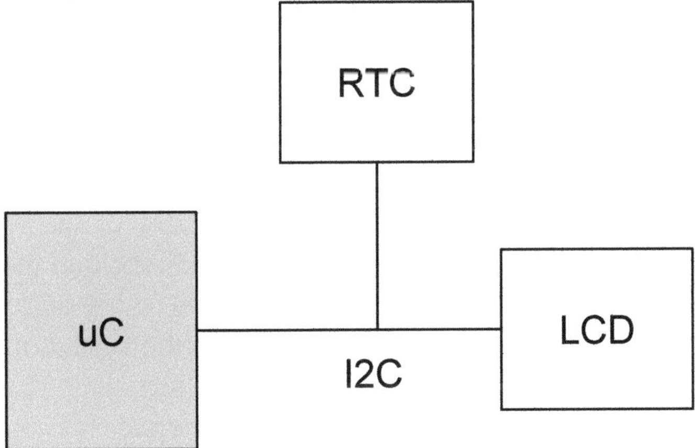

Figure 31 *Multi-slave I2C configuration*

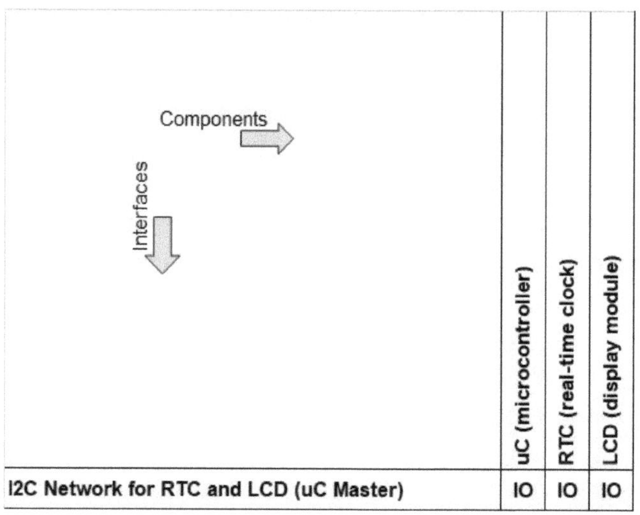

Figure 32 *I-CM of the multi-slave I2C configuration*

Our uC supports fully software-configurable I2C. This means that the high and low signal levels could be configured by the software, as well as the clock speed. The RTC and LCD components, however, don't have this flexibility. If the three of them share the same network, then there should be a common configuration between them. But beware - knowing that there is communication only between master and slave, ensuring compatibility of the master and each of the slaves is not enough, as every slave could "listen" to every communication on the bus. If the logical levels of the RTC are perfectly aligned with the uC, but are on the threshold of the trigger of the LCD, the messages sent from the RTC to the uC could be misinterpreted by the LCD in many ways, and the system may start behaving unpredictably (imagine if the 0s were sometimes perceived as 1s and vice-versa, which may generate all kinds of miscommunication). Then for this interface, we may consider the following design and verification attributes:

DESIGN:
- Select components with compatible I2C clock speeds. Ensure that the common speed enables enough bandwidth for the application.
- Communicate the software configuration required for the microcontroller (as the master of the I2C) to the software team.
- Consider signal levels of compatibility of each device on the bus with every other device (each with each, not only each slave with the master!). Consider serial resistors to unify the signal levels for the devices with higher Vcc if needed.

VERIFICATION:
- Measure the communication with each device-controlled message. Confirm the signal levels compatibility of each device with every other, even when not intended to communicate with each other.
- Verify the signal levels at Tmin and Tmax and confirm the bus compatibility is kept within the operational range.
- Run stress-testing on maximal network load and confirm the clock speed selection.

Why is it a system interface? Because it is shared between EE and software.

Should it be addressed in the system architecture? Yes. Additionally, the microcontroller master configuration should be derived from the system architecture and described in detail in the software/hardware interface specification.

end of the example

All the technical details we see in those examples may be scary at first. There are so many things to be considered for the system architecture, and so much detailed verification to be done during the integration - who are those super-engineers with such deep, detailed understanding of all disciplines? But refer again to Figure 13 - the systems engineers may involve specialists during each activity and ask: What needs to be considered for the design of the solution? In this way, the input for the I2C software configuration, the serial resistors, and such, may come from an EE Senior Engineer. Later, the EE team implementing the design may not be as experienced, but the considerations of how to design the solution (given by the Senior Engineer as input) will be propagated via the system architecture for their knowledge. The detailed measurements of the signals on the I2C lines with an oscilloscope or signal analyzer could be performed by the EE team on request of the systems engineer, and the results could be taken as an integration test result. The systems engineer is not free of "doing" but is also not alone when addressing the complexity of the system.

> *The systems engineer may delegate tasks and collect recommendations and results from all disciplines when detailed work is required.*

Final words about the system integration and system testing - keep in mind that most failures come from interfaces. Another source could be the dimensioning of the components - they may work perfectly in the simulated environment, but may fail as soon as the system is put together. Test every interface and don't skip the corner cases. Load the system and get the critical system-relevant metrics of the components (reaction time, processor load, memory consumption, different utilizations, etc.). During the integration, a lot of insights for the system will come due to the white-box nature of the testing. Take advantage of this!

Validate the Result (System Testing)

In the previous chapter, we assembled our system and verified that it works right - this means - what we got is what was intended by design. Initially, the design was made in order to resolve our initial problem. In theory, if the design resolves the problem and the system is built as per design, then the system should resolve the initial problem as well, right? In the real world, this is virtually never the case (and even if it is, we would like to be sure).

System testing (also called validation) aims to check if the final system is really the solution to the initial problem. If you remember refining the inputs for the system architecture (Figure 20) - this is what defined our solution. In this sequence of user requirements -> system requirements -> functional model, we might have missed something somewhere. System testing can use any of those inputs to define its testing scenarios, but everything should be traced down to the user requirements at the end.

> *The ultimate goal of system testing is to prove that the system is a solution to the initially-stated problem or need.*

The test plan should cover all the requirements, and for those which are not testable, there should be another way to verify them - with review, assessment, statistics - whatever works.

In the whole process at the left side of the "V" (Figure 12A), the main use cases of our system should have been defined. These should be executed during the system testing as well, driving each of the use cases to its extremities, in addition to the nominal case. For example, if a product is designed for users with weight between 40 and 120 kg, all the test cases where the users and their weight may play a role should be tested with a nominal weight (say 80 kg), plus the extremes 40 kg and 120 kg. Pushing the boundaries until failure is an additional approach to assess the robustness of the solution - in the case of the weight of the user, this would be increasing the weight gradually (say, in steps of 10 kg) until the product fails. Generally, system testing creates a controlled system context and verifies how the system reacts to it, while delivering the solution to our initial problem. System testing should not try to prove the compliance of the product with the architecture or the concept (we already did this with integration testing). The system at this stage is already a closed box (black box) and it is tested as it would be used in the real world.

The test plan creation is usually done in parallel to the system architecture activity. The requirements specification is enough to start creating our test scenarios. If we create a test plan **knowing** the solution and adapting our test cases to it, we are biased and may go to great lengths to confirm that a great system was created... but we wouldn't know if this system is what is needed.
The scope of the test plan should not be limited only to the product-specific requirements. We may include all relevant generic, legal, or product-type requirements in our test plan - like electromagnetic emissions, electrostatic discharge resistance, chemical emissions (of both the final product and the production process), shock resistance, lifetime requirements (ability to operate in high and low humidity, high and low temperature, under vibration and mechanical shock, and within the lifetime expectancy), etc. For many industries, these are specified as common "norms" and may be used by the organization as generic product requirements.

The full contents of the test plan may look huge, but full execution is not always necessary. If we have a product development plan in which we release new hardware twice over the whole development cycle, but new software 20 times - we will perform the strictly hardware-related tests (like lifetime, shock, vibration, EMC, ESD, etc.) only on a new hardware release. <u>A tricky part:</u> In many cases, the software configures the physical characteristics of electrical/data signals of the hardware (shape, frequency, amplitude, etc). These may have an impact on the electromagnetic emissions or signal integrity, for example. Configurable threshold settings for thermal protection of microcontrollers may cause different behavior in high-temperature working conditions, etc. A good impact analysis of the software changes may show a need for delta testing of the hardware behaviour too, even though no new hardware has been released!

It is also not necessary to perform full functional validation with every software release - if the changes to the previous software release and their potential side impacts are evaluated, a delta validation could be done over only the changed *and* impacted features.

These decisions should be documented in a test plan strategy, where the scope of each test campaign is defined and justified.

In the I-CM tool, the rough functional test plan is contained as the sum of the "verification" attributes of the system's requirements / functions (although strictly speaking it is "validation" from system perspective).

As an example, let's take the Vehicle's E/E model on Figure 26 and the selected feature *display fuel level*. Some preliminary information, before we start with the test plan: The tank sensors send over the LIN once per second a value between 0 and 100, indicating the percentage of fuel in the tank. The instrument cluster gets the message over the Multimedia CAN (via the gateway) and displays it on the screen as an analog indication (pointer on a scale). The tank has a capacity of 50 L.

The test plan should describe the prerequisites for each test case, the execution steps, and the expected results. Of course, each test case should be linked to a requirement(s)/feature. Now, we know that all the following test cases are linked to our high-level feature *display fuel level*:

Display Fuel Level - Test Plan:

Remark: Test the instrument cluster functionality

1.   <u>Prerequisites:</u> Instrument cluster, simulation connected to the Multimedia CAN

<u>Test Sequence:</u> Send CAN messages for tank indication of 0, 25, 50, 75, and 100%

<u>Expected Results</u>: Indication of 0, 25, 50, 75, and 100%

2.   <u>Prerequisites:</u> Instrument cluster, simulation connected to the Multimedia CAN

<u>Test Sequence:</u> Send CAN messages for tank indication of random negative value and random value more than 100

<u>Expected Results:</u> Indication of 0 and 100%

Remark: Test the message path from the LIN to the instrument cluster

3.   <u>Prerequisites:</u> Instrument cluster, gateway, simulation connected to the LIN

<u>Test Sequence:</u> Send LIN messages for tank indication of 0, 25, 50, 75, and 100%

    <u>Expected Results:</u> Indication of 0, 25, 50, 75, and 100%

4.   <u>Prerequisites:</u> Instrument cluster, gateway, simulation connected to the LIN

    <u>Test Sequence:</u> Send LIN messages for tank indication of random negative value and random value more than 100

    <u>Expected Results:</u> Indication of 0 and 100%

Remark: Test the Vehicle-level function and accuracy of indication depending on the Vehicle position

5.   <u>Prerequisites:</u> Instrument cluster, gateway, tank sensors, tank, Vehicle parked on a flat horizontal surface

    <u>Test Sequence:</u> Fill the tank with 0, 12.5, 25, 37.5, and 50 L of fluid

    <u>Expected Results:</u> Indication of 0, 25, 50, 75, and 100%

6.   <u>Prerequisites:</u> Instrument cluster, gateway, tank sensors, tank filled with 25 L of fluid. Vehicle on an adjustable stand.

    <u>Test Sequence:</u> Set the stand surface at 10° - down to the front, rear, left, and right of the vehicle (4 positions)

    <u>Expected Results:</u> Indication of 50% +/- 5% at each position

Of course, we could start directly with test case No. 5, but if it fails, we will not know where the problem may be. If test case No. 5 fails, after all prior test cases have passed, the issue should be in the tank sensors (or the tank may have been produced with the wrong geometry or volume). While debugging the issue is not directly in the scope of system testing, some additional localization included in the test report can only help.

For example, if test case No. 5 fails, the following additional tests may be done:

- Read the message directly on the LIN network and compare it with the indication on the instrument cluster (to see if the network communication of the tank sensors failed).
- Fill the tank with 20 L and then add 10 L. Is the second indication 150% of the first one? If yes - the physical to relative mapping at tank sensors is wrong **or** the volume of the tank is wrong.
- And so on...

In any case, it is much easier to identify the source of an issue if we perform incremental testing, increasing or localizing the scope with each following test case.

After finishing a test campaign, the results (test report) should be stored and baselined (to track which test report corresponds to which product release). For each detected failure or non-conformity, a defect ticket should be opened. However, this belongs to other processes, which presumably organizations already have - configuration management, defect management, and product release processes.

Introduction to the I-CM Method

The I-CM (Interface-Component Model) is a simple method which aims to cover the fundamental aspects of a system in an integrated view (Figure 33).

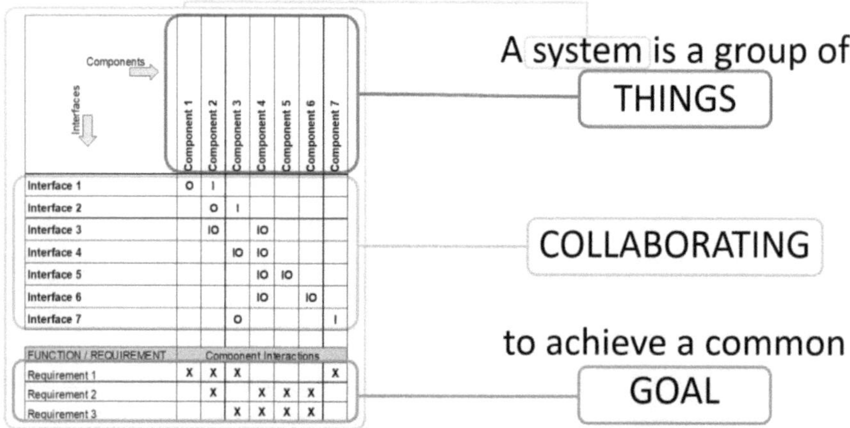

Figure 33 *The I-CM represents the system as per definition*

The I-CM is a dual matrix. The physical system is represented as a map between *interfaces* and *components* (or "interactions" and "things") - this is the I-CM core (Interface-Component) and the origin of its name. On the *component* dimension, another link is established between the components and the *functions/requirements*, building an adjacent matrix to the first one to display how the physical solution satisfies the requirements ("goals"). In this way, we have a combined view of the key aspects of a system as per the "system" definition: the "things," the "collaboration," and the "goal."

The I-CM forces us to treat the interfaces and the components equally. The interface is not simply defined as input/output of two (or more) components (Figure 34), or as the intersection between two components[3] (Figure 35).

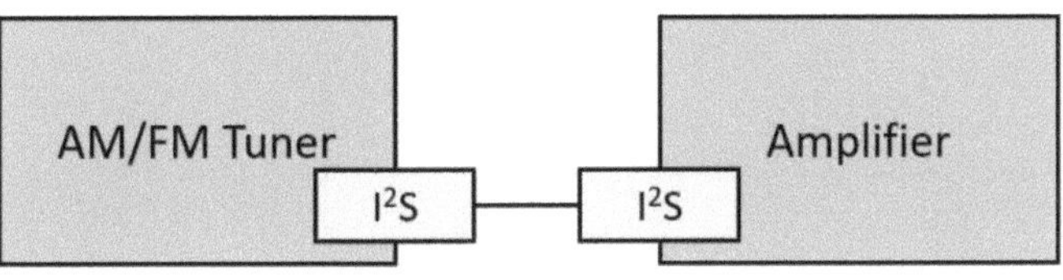

Figure 34 *I2S interface between two nodes*

	Component 1	Component 2	Component 3	Component 4	Component 5
Component 1			X	X	
Component 2					X
Component 3	X				
Component 4	X	X			
Component 5			X		

Figure 35 *DSM example*

Instead, the interface in I-CM is treated as an item, just like the components, which has its own identity and purpose, design specifics and properties, and verification criteria (the I-CM may be extended, if needed, by adding even more fields, like - owner, plan,

---

[3] *The example is a DSM - an excellent tool with a specific purpose, which shouldn't be considered interchangeable with I-CM.*

progress, etc). Figure 34A shows the core matrix of the I-CM, representing two components with one interface between them.

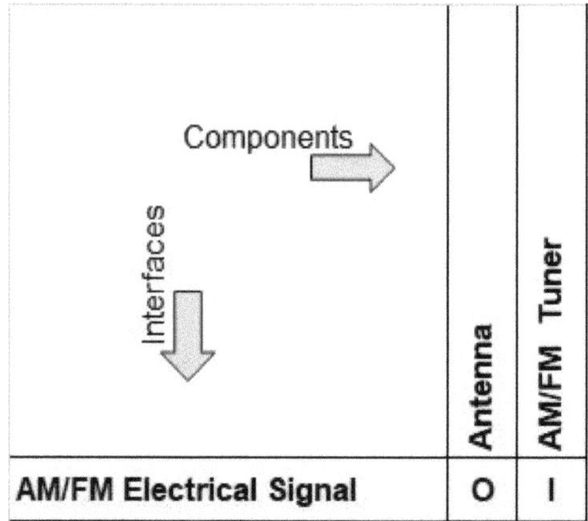

Figure 34A *The notation "I" and "O" mean "Input" and "Output" in the table.*

Let's do some practice with the I-CM using the following example: an AM/FM Receiver with digital controls. It should receive AM and FM radio signals and play the sound of the selected station through the speaker. The selection of the band, radio station, and volume is done by the user via keyboard, and displayed on an LCD. Let's define the following high-level functional requirements for this product (as you see, these are neither atomic nor abstract, i.e., they are user requirements):

- **REQ #01**: Receive an AM/FM signal from a selected band/station and play the sound with the selected volume through the speaker.
- **REQ #02**: Allow the user to select the band and the station via keyboard and display it on an LCD.

- **REQ #03**: Allow the user to select the audio volume via keyboard and display it on an LCD.

Figure 36 shows a possible solution for our system, represented in a diagram (EE view, power domain not included).

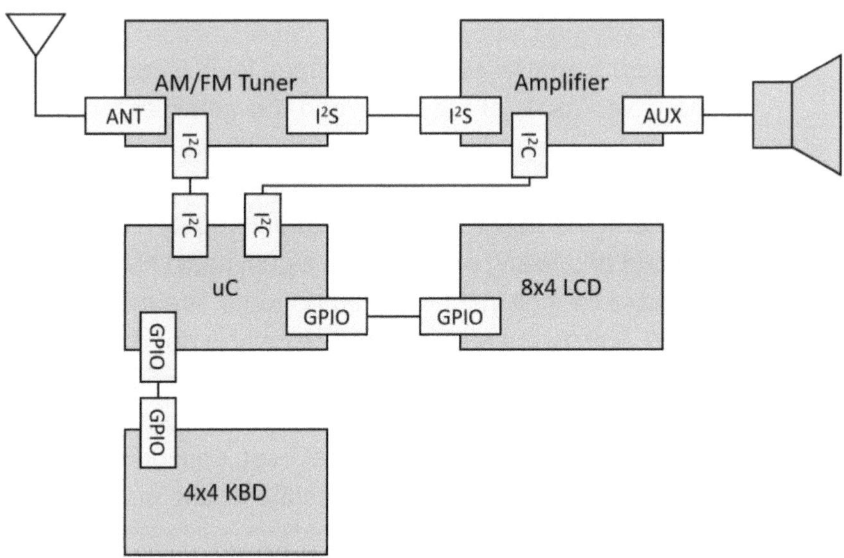

Figure 36 *Abstract E/E architecture of a radio receiver*

Let's do some thinking about the I2S interface between the components AM/FM tuner and amplifier now:
- I2S is a three-line bus. The two components (which are physically two ICs) need to be connected via three physical electrical lines. Possibly PCB traces and not wires?
- How will these traces be rooted (length, form, in which layers of PCB) in order to avoid them acting as antennas for EM emissions?
- One of the I2S lines is a clock, which needs to be configured via software. However, CD audio quality is calculated to be

89

~1.41MHz, and this frequency is in the AM/MW range. How will we prevent potential injection of this signal to the AM/FM tuner antenna input? Place the two components very near to each other to have a short trace? Shield the I2C traces between PCB layers (although there may not be enough PCB layers in such a simple application)? Use a different clock frequency?

- The trigger signal levels (which voltage level is recognized as 0 and which as 1) of the tuner and the amplifier for the I2S interface lines may be different. Should we consider serial resistors on the lines to unify the levels? What would be the influence of the resistors on the signal integrity?

- The rising and falling edges of the signal could have different acceptable timings in the two components. Are they common ranges? A software configuration usually is needed to set up these settings to be compatible for both I2S nodes. What is the influence of the variations in the signal edges on the EMC performance, what would be the best configuration for a stable signal, and which has minimum EM emission and maximum EM immunity?

- And so on…

All the design decisions based on questions like this, including their verification plans, should be documented in the I-CM for the entry of this I2S interface. Some of them may be propagated to the implementation of the involved components.

From an organization point of view, very often the ownership of the interface is not clear, and when one designer works on the tuner and another works on the amplifier block, no one actually works on the "line" between them. In addition, the interface has a purely interdisciplinary nature (and in this sense is not only EE, but system-relevant), as it is not only in the EE designers' control because the PCB layout engineers, environmental testers, and software engineers use or contribute to this interface.

Now, let's represent the same system in the I-CM (Figure 37). The relations between the components over the interfaces may be:
- Input, **O**utput and **IO** (for any flow of information, energy, material),
- But also **M**echanical, **S**tructural, **N**egative, etc. (except **I** and **O**, every other notation is considered bi-directional).

We also include the mapping between the "function" and the "form" in the second matrix, where the increasing numbering defines a "path" or execution sequence. Parallelism or asynchronous interactions are represented with the same number, indicating no sequencing.

Components → / Interfaces ↓	Antena	AM/FM Tuner	Digital Amplifier	microController	LCD	4x4 Keyboard	Speaker		:Interface connectivity
AM/FM Electrical Signal	O	I							2
I2S Audio Data Interface		O	I						2
I2C command line tuner		IO		IO					2
I2C command line amplifier			IO	IO					2
8xGPIO for LCD control				IO	IO				2
8XGPIO for KBD input				IO		IO			2
Analog Audio Signal			O				I		2
AM/FM Radio Signal	I								1
									0
Component connectivity:	2	3	3	4	1	1	1	0	

FUNCTION / REQUIREMENT	Component Interactions								DIST
Receive AM(535-1605 kHz)/FM(66-108MHz) radio signal and play the sound with selected volume(0-3W) over the speaker	1	2	3				4		4
Allow the user to select the band and the station via keyboard and display it on an LCD;		1		1	1	1			4
Allow the user to select the audio volume via keyboard and display it on an LCD;			1	1	1	1			4
									0
Functional Load	1	2	2	2	2	2	1	0	

Figure 37 *The I-CM of the radio receiver*

What we can see so far in this representation:

- List of interfaces and components and how they are related to each other.
- Level of connectivity of each component, where 0 means a component is not part of the system and a higher number indicates high dependency in the system (here, the microcontroller is the highest).
- Connectivity level of each interface: 0 means the interface is irrelevant to the system, 2 means peer-to-peer interface, more than 2 means the interface is a kind of bus, multi-connection point in mechanics, etc. A connectivity level of 1 may be used as a shortcut to indicate an interface to an external component - if we add the AM/FM radio signal as an interface, but don't add the external sender, then we may connect this interface as an Input to the antenna only.
- We can see the complete subsystem delivering a function, and in some cases also the sequence. For example, radio reception (REQ #01) - we can see not only the sequence of components involved to deliver the function, but also which interfaces it goes through (Figure 38).

Interfaces \ Components	Antenna	AM/FM Tuner	Digital Amplifier	microController	LCD	4x4 Keyboard	Speaker	:Interface Connectivity
AM/FM Electrical Signal	O	I						2
I2S Audio Data Interface		O	I					2
I2C command line tuner		IO		IO				2
I2C command line amplifier			IO	IO				2
8xGPIO for LCD control				IO	IO			2
8XGPIO for KBD input				IO		IO		2
Analog Audio Signal			O				I	2
Component connectivity:	1	3	3	4	1	1	1	

FUNCTION / REQUIREMENT	Component Interactions							
REQ#01	1	2	3				4	4
REQ#02		0		0	0	0		4
REQ#03			0	0	0	0		4

Figure 38 *Subsystem highlight in the I-CM*

To make the I-CM even more practical, there are two additional documentation properties (implemented in the spreadsheet tool as foldable fields). Following the philosophy of the I-CM method to treat interfaces and components equally, these properties are applicable for both of them. The properties are used to document the design decisions as well as the verification plan for the component or interface. The design decisions are propagated to the team responsible for the respective implementation. The verification plans of the components are usually executed during the component verification (often outside of the system), while the interface verification is executed in the integrated system during the integration testing. In both verification cases, instrumentation can be utilized for the sake of thoughtfulness. A good approach to designate an interface owner is to assign it to the master component using it. The tool could be extended with additional properties for the components, interfaces and requirements, like "owner," "status," "target date," and so on (Figure 39).

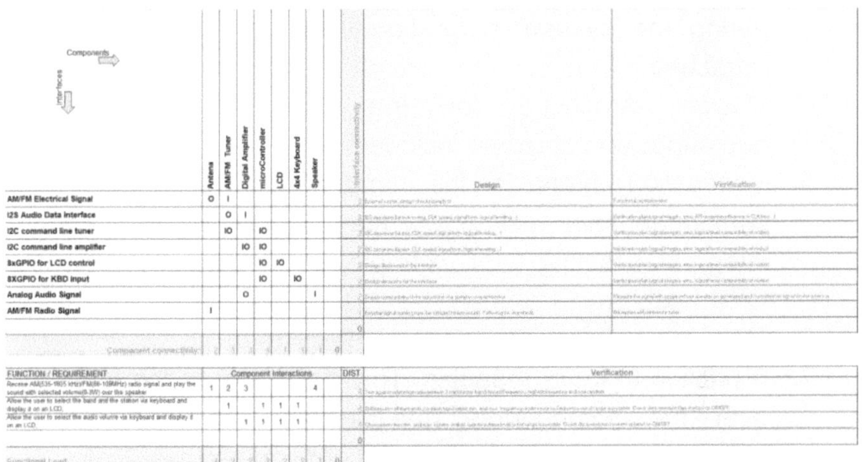

Figure 39 *I-CM with document properties for design and verification*

To summarize again the content of the I-CM and how it covers the basic needs of the SE activities from the chapter "Introduction to Systems Engineering" (Figure 40):

1. The *physical system* is represented as components and interfaces (or things and interactions). This is the so-called I-CM core matrix.
2. *Design* directions for the interfaces. These could be propagated to the subsystems in the design, or to requirements of the related components. Integration prescriptions should be included as well.
3. *Verification* criteria for the interfaces. These should be tested during the system integration with the real components "attached."
4. *Design* directions for the components. Discipline decomposition could be included.
5. *Verification* criteria for the components. This is mostly for component-level testing, but may include requirements for integration testing - for example, performance metrics like CPU load.
6. *Requirements* and *functions* of the system.
7. "Function to form" mapping. This decomposes the functions over the components (the related interfaces will be automatically linked via the I-CM core matrix).
8. *Functional test* plan.

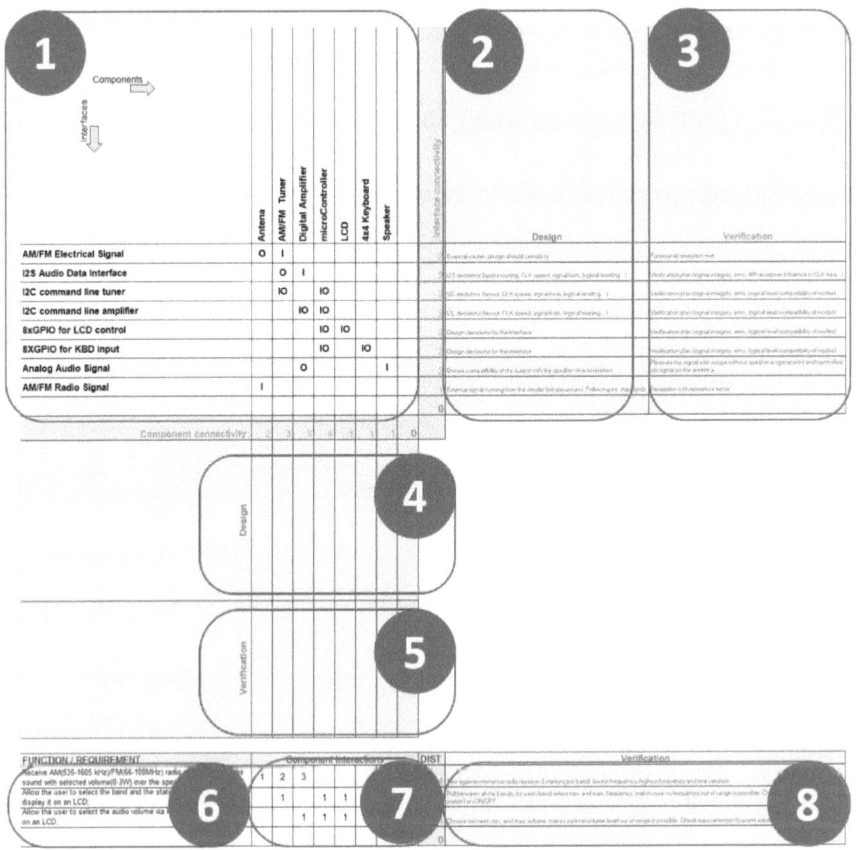

Figure 40 *I-CM 1.x content overview*

Quick Reference to I-CM 1.x

The I-CM method may be used as a simple pen-and-paper tool, but computer implementation is much more powerful. The following is a quick reference to the current version of the I-CM (ver. 1.x) for MS Excel™.

To add a new system element - component, interface, requirement, or function - simply start typing in the **last** free cell of the respective list (Figure A1). **Do not** use the Excel functions "Insert rows" or "Insert columns" as this will corrupt the boundaries tracking of the scripts!

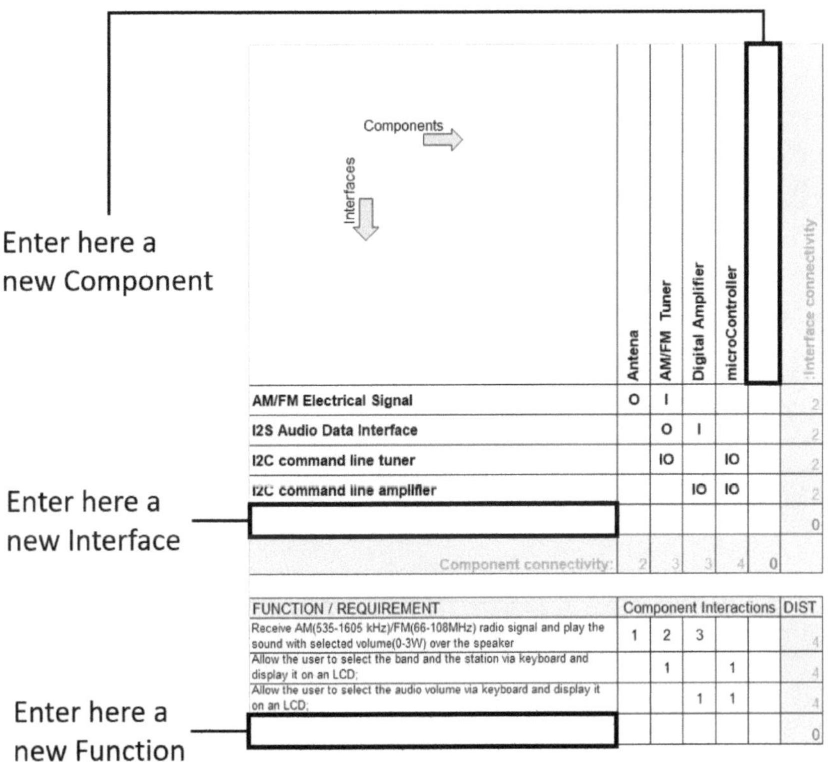

Figure A1 _Entering a new system element in the I-CM_

Removing an interface, component, or requirement

To remove any component, interface, or requirement/function, just delete its name (select the cell and press DEL button), and you will be prompted to proceed with deletion of this system element from the system (Figure A2). **Do not** use Excel functions "Delete" or "Cut"!

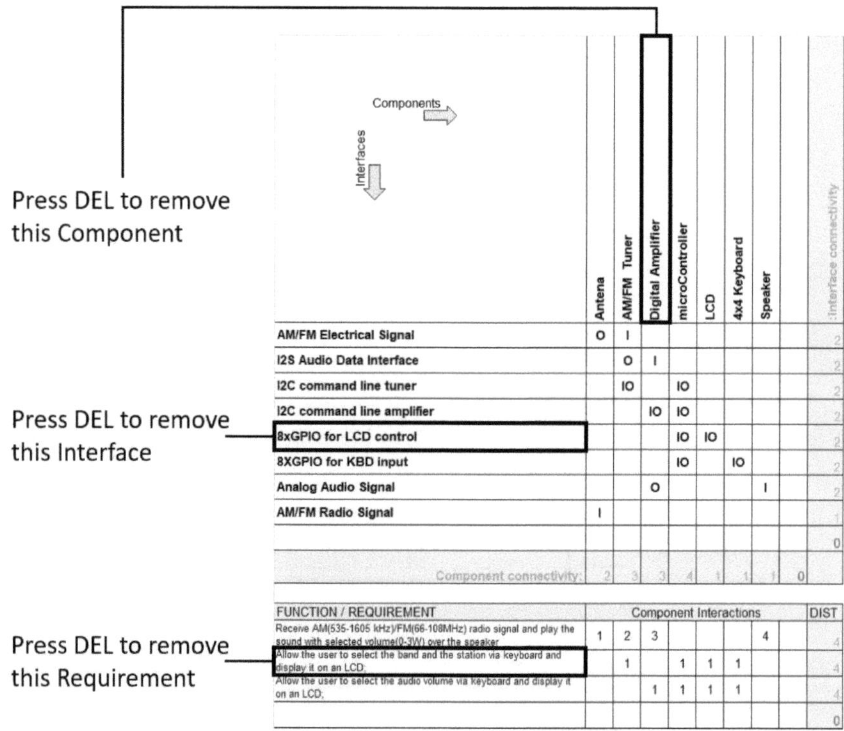

Figure A2 *Removing a system element*

In the I-CM there are two types of relationships - between interfaces and components (once established, they will define the physical system), and between components and functions (mapping form to function).

To establish a relationship between components, an interface should be defined first (Figure A3). Once the components **and** the interface are present in the I-CM core matrix, enter a letter in the intersection cells between the interface and the related components. Predefined letter(s) are:

- **I** - Incoming information / energy / material flow
- **O** - Outgoing information / energy / material flow
- **IO** - Incoming and outgoing
- **M** - Mechanical (snap-on, solder, welding, glue…)
- **S** - Static connection
- Any other letter could be used to indicate a custom relation (including an unwanted/negative relation). Other letters will be interpred like I/O for the DSM generation.

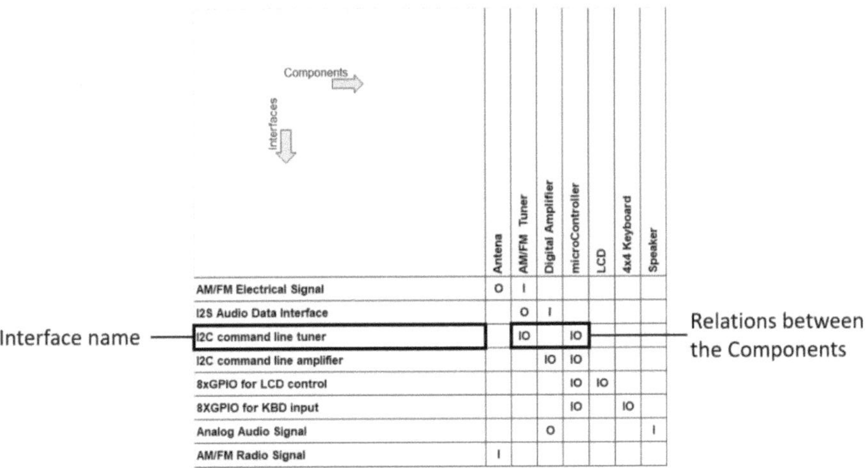

	Antena	AM/FM Tuner	Digital Amplifier	microController	LCD	4x4 Keyboard	Speaker
AM/FM Electrical Signal	O	I					
I2S Audio Data Interface		O	I				
I2C command line tuner		IO		IO			
I2C command line amplifier			IO	IO			
8xGPIO for LCD control				IO	IO		
8XGPIO for KBD input				IO		IO	
Analog Audio Signal			O				I
AM/FM Radio Signal	I						

Figure A3 *Interactions definition*

To establish a relationship between the functions/requirements and physical subsystems, in the lower matrix, in the intersection cells between the function and the related components, enter a **number** to define a relation (Figure A4). The usage of numbers gives the flexibility to document a sequence or path of execution. Aside from this, the numbers do not bear any weight; they only establish a link between the function and a component.

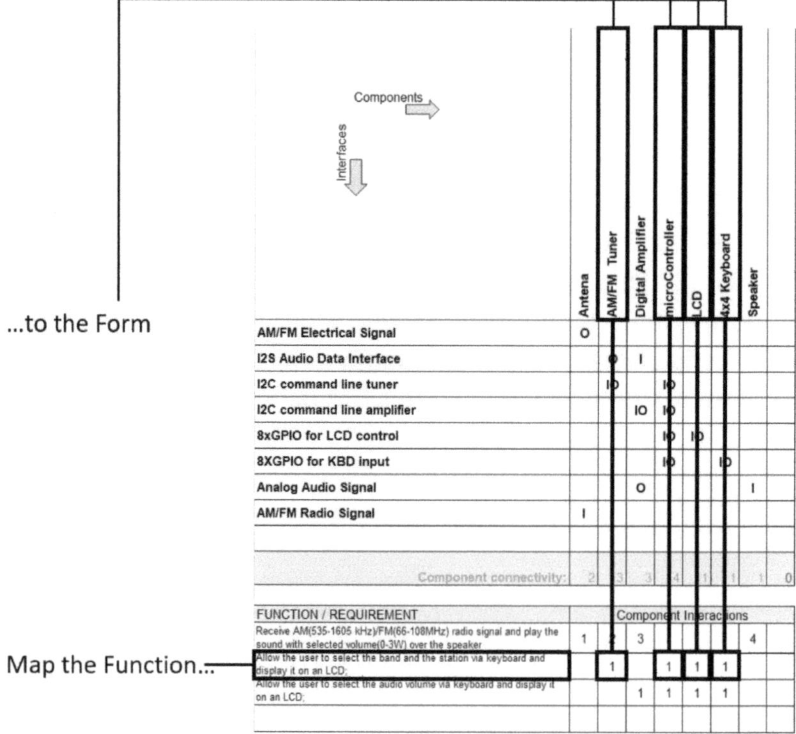

Figure A4 *Map form to function in the I-CM*

The System Highlight feature of the I-CM tool is very useful, enabling quick visual exploration of the relations within the system. Once a system element (interface, component, or function) is selected, the related subsystem is automatically highlighted.

If a component is selected, all its interfaces and the functions dependent on the component are highlighted (Figure A5). This could be used for impact analysis, incremental integration step definition, test planning, implementation planning, etc.

Interfaces \ Components	Antena	AM/FM Tuner	Digital Amplifier	microController	LCD	4x4 Keyboard	Speaker	
AM/FM Electrical Signal	O	I						
I2S Audio Data Interface		O	I					
I2C command line tuner		IO		IO				
I2C command line amplifier			IO	IO				
8xGPIO for LCD control				IO	IO			
8XGPIO for KBD input				IO		IO		
Analog Audio Signal			O				I	
AM/FM Radio Signal	I							
Component connectivity:	2	3	3	4	1	1	1	0

FUNCTION / REQUIREMENT	Component Interactions							
Receive AM(535-1605 kHz)/FM(66-108MHz) radio signal and play the sound with selected volume(0-3W) over the speaker	1	2	3				4	
Allow the user to select the band and the station via keyboard and display it on an LCD;		1		1	1	1		
Allow the user to select the audio volume via keyboard and display it on an LCD;			1	1	1	1		
Functional Load	1	2	2	2	2	2	1	0

Figure A5 *Component impact on the system*

Selecting an interface will highlight all the components interacting via the interface (Figure A6). This is useful for incremental integration step definition, dependency analysis, and implementation planning.

104

	Antena	AM/FM Tuner	Digital Amplifier	microController	LCD	4x4 Keyboard	Speaker	
AM/FM Electrical Signal	O	I						
I2S Audio Data Interface		O	I					
I2C command line tuner		IO		IO				
I2C command line amplifier			IO	IO				
8xGPIO for LCD control				IO	IO			
8XGPIO for KBD input				IO		IO		
Analog Audio Signal			O				I	
AM/FM Radio Signal	I							

Figure A6 *Interface highlight*

The selected function/requirement will highlight the physical subsystem (known as the minimal system) which delivers this function (Figure A7). In addition to defining minimal systems, this is very useful for performing functional impact analysis, subsystem integration definition and testing, and feature-based planning.

Interfaces \\ Components	Antena	AM/FM Tuner	Digital Amplifier	microController	LCD	4x4 Keyboard	Speaker	
AM/FM Electrical Signal	O	I						
I2S Audio Data Interface		O	I					
I2C command line tuner		IO		IO				
I2C command line amplifier			IO	IO				
8xGPIO for LCD control				IO	IO			
8XGPIO for KBD input				IO		IO		
Analog Audio Signal			O				I	
AM/FM Radio Signal	I							
Component connectivity:	2	3	3	4	1	1	1	0

FUNCTION / REQUIREMENT	Component Interactions							
Receive AM(535-1605 kHz)/FM(66-108MHz) radio signal and play the sound with selected volume(0-3W) over the speaker	1	2	3				4	
Allow the user to select the band and the station via keyboard and display it on an LCD;		1		1	1	1		
Allow the user to select the audio volume via keyboard and display it on an LCD;			1	1	1	1		

Figure A7 *Highlighting functional impact and minimal system*

Documenting the interfaces

Every interface entry consists of the Name (explaining main purpose), relationships to the components, Design, and Verification (Figure 40). The fields Design and Verification are foldable to allow easier viewing of the system. Refer to the previous chapter about this.

Documenting the components

Every component entry consists of the Name (explaining main purpose), relationships via interfaces, Design, and Verification (Figure 40). The fields Design and Verification are foldable to allow easier viewing of the system. Refer to the previous chapter about this.

Documenting the requirements

The functions/requirements are either referred to via ID, or the whole text description is entered in the first cell of the "form to function" mapping matrix. Then the mapping itself is established with the components (physical part of the system), and at the end, the plan to test the function is documented in the Verification field (Figure 40).

Generate DSM from the I-CM system model

The DSM (Design Structure Matrix - https://www.dsmweb.org/) is a powerful tool to analyze and cluster complex systems. The I-CM contains enough information to generate a DSM out of the system model in the core matrix of the I-CM. Every time the "DSM" tab of the I-CM tool is opened, a new DSM matrix is generated out of the I-C model. This could be transferred to another DSM tool for additional processing. Import from DSM is not optimal, as there will be loss of information back to the I-CM, but the DSM optimizations could be insightful for a systems engineer.

Abbreviations

BOM   - Bill of Materials

CAN   - Controller Area Network
CPU   - Central Processing Unit

DSM   - Design-Structure Matrix
DSP   - Digital Signal Processing

ECU   - Electronic Control Unit
EE   - Electrical and Electronics
EM   - Electro-Magnetic
EMC   - Electro-Magnetic Compatibility
ESD   - Electro-Static Discharge

FMEA - Failure Mode and Effect Analysis

IC   - Integrated Circuit
I-CM - Interface-Component Model

LCD   - Liquid Crystal Display
LIN   - Local Interconnect Network

MBSE - Model-Based Systems Engineering
ME   - Mechanics

OEM - Original Equipment Manufacturer
OSP   - Organic Solderability Preservative

PC   - Personal Computer
PCB   - Printed Circuit Board

RTC   - Real-Time Clock

SE   - Systems Engineering (or Systems Engineer, depending on
         the context)
SMD   - Surface-Mounted Device
SoS   - System of Systems
SW   - Software

Bibliography

Crawley, Cameron, and Selva. *System Architecture - Strategy and Product Development for Complex Systems.* Pearson, 2016.

Ulrich, Eppinger. *Product Design and Development (Sixth Edition).* McGraw Hill, 2016.

Reinertsen. *Managing the Design Factory - a Product Development Toolkit.* The Free Press, 1997.

Bahill, Dean. *What is Systems Engineering? A Consensus of Senior Systems Engineers.* Sandia National Laboratories, 1995.

Gausemeier, Dumitrescu, Steffen, Czaja, Wiederkehr, Tschirner. *SYSTEMS ENGINEERING in industrial practice.* University of Paderborn, 2015.

DoD - Systems Management College. *Systems Engineering Fundamentals,* 2001.

NASA. *Systems Engineering Handbook - NASA/SP-2007-6105 Rev1,* 2007.

VDA QMC Working Group 13/Automotive SIG. *Automotive SPICE Process Assessment / Reference Model.* 2015.

Tuzsuzov. *System Architecture:Applying the Lessons Learned in the Early Design Stages of Visteon Engineering Process.* Visteon Corp. WhitePaper, 2019.

Tuzsuzov. *Holistic Engineering: Interface-Component Model (I-CM).* WhitePaper https://people.rit.edu/~yt3007/, 2020.

<u>Lectures and classes:</u>

Massachusetts Institute of Technology (MIT xPro), *"Architecture and Systems Engineering: Models and Methods to Manage Complex Systems"*

Rochester Institute of Technology (RIT), *"Engineering of Systems - I"* ISEE-771

ABOUT THE BOOK:

This book is a hands-on introduction to the basic concepts of systems engineering. The various examples, used to illustrate each of the discussed topics, help the reader to understand the concepts more easily. The book presents a simple method called the I-CM (Interface-Component Model), which enables practical implementation when no other tools are available.

"Systems Engineering for All" is intended for a general public of engineers and product designers without prior systems engineering experience. It is not an academic book.